RED PLAGUE
BLACK DEATH

BY

TONY WALTON

This book is a work of non-fiction. Names and places have been changed to protect the privacy of all individuals. The events and situations are true.

ISBN: 1-4107-8732-X (e-book)
ISBN: 1-4107-8733-8 (Paperback)

This book is printed on acid free paper.

Front cover designed by Ashley and Tara Walton

1stBooks – rev. 09/15/03

Dedicated to my wife Charlene who told me to stop talking about writing a book and to just go and do it.

Introduction

The wave of change sweeping across Africa has finally reached South Africa. An economically and politically dominant white minority had for decades ruled this mineral rich country, with a total population of more than thirty million people. This minority of five million, descendants of the original Dutch and English settlers, is still coming to terms with the new reality of equality for all citizens of South Africa.

Prior to 1994, when the previously banned African National Congress came to power, South Africa had only known decades of white rule under the *apartheid* policy. At the heart of this policy was the 'Group Areas Act,' a government dictate that was designed to reduce the number of 'non whites' in what had been officially designated as 'White South Africa'.

This act enforced 'pass laws' that strictly monitored the movement of all 'non whites.' It also removed from the large affluent centers of Johannesburg, Cape Town and Durban all black or colored Africans not considered economically necessary, laborers, domestic servants, and mine workers.

In the years 1966 to 1975 prosecutions under laws restricting the movement of black and colored South Africans numbered more than 5.8 million persons. Black opposition to this discriminatory rule took the form of mass protest and civil disobedience. Then on March 28 1960, in a black township called Sharpeville, a demonstration against the extension of pass laws to include black women, finally resulted in a violent confrontation. The police killed 69 people and wounded 178.

The African National Congress responded to the massacre by calling a General Strike, which paralyzed the country for three weeks. The ruling Nationalist government then retaliated by declaring a state of emergency and detained more than ten thousand of the dissidents, without trial.

News of the Sharpeville massacre and the ensuing government crackdown brought international condemnation and South Africa now found itself diplomatically and economically isolated. Despite this grave threat to the economic growth of the country, together with a huge outflow of internal capital, South Africa continued to enforce its undemocratic and vilified policy.

It was not until the free and democratic elections of April 1994 and the appointment of Nelson Mandela as President that South Africa was at last able to throw off the yoke of apartheid.

Despite these recent changes, the scars from decades of racial intolerance and exploitation have still to be healed and some whites have yet to accept the reality of a less privileged status in the new South Africa.

There are still reactionary forces at work, bent on undermining the fledgling democratic process and who seek to return power to the white minority.

What now follows is the story of a depraved scientist who will stop at nothing, murder…sabotage…even genocide, to gain power over his fellow countrymen. The story is fictional but it is based on scientific fact.

Chapter 1:

Johannesburg: South Africa

The female mosquito was thirsty for blood. Her male brethren were content to feed on nectar or water but her mouth parts were longer, adapted over millions of years so they could pierce flesh and inject salivary fluid into the wound and then draw up the red fluid she craved. Like other females of her species, she was able to drink about twice her weight in blood at one time, for without this precious red nectar she was unable to lay her eggs. Unlike mosquitoes in colder climates, she could drink blood and lay eggs all year round. She could also transmit malaria, encephalitis and dengue fever.

Trimming her gossamer wings she glided on the warm air currents in preparation for her attack and then started her descent onto one of the many luxurious houses below. It was a hot dry morning, which was unusual for Johannesburg so early in September, and the bright African sun had already begun to beat down relentlessly on the affluent suburbs that stood out so proudly in this northern sector of Southern Africa's wealthiest and largest city.

The airborne blood-sucker was oblivious to the changes that had recently occurred in this once white dominated country. The newly liberated black tribes now freed from oppression by the democratic government of Nelson Mandela were of no interest to her. Her only concern was for the taste of blood, blood from some warm host, animal or human, black or white.

Reaching ground level the thirsty insect headed for the open windows of a nearby house where strong beams of sunlight were shafting through the large picture windows that framed this particular villa. A man who was lounging inside turned his muscular body to absorb the familiar heat that was permeating his already deeply tanned body.

Commandant Marius Johannes Joubert scoured the paper that lay on the table in front of him, scanning the headlines casually, until suddenly his attention was drawn to a minor article on the back page. As his still waking mind absorbed the contents of the story he suddenly flung the newspaper across the room with a loud snort of disgust. The newspaper hovered helplessly in mid air before it disintegrated into a crumpled heap, the offending article still mocking him from the floor. *Deadly Virus breaks out in Congo*, the headline screamed.

The article went on to describe in detail that a team of American Virologists had just jetted out from a special investigation unit in Washington, to help with the suspected Ebola outbreak, now reported by a small village sixty miles West of the capital of Kinshasa.

Joubert was a tall, powerfully built man in his late forties, with thick black hair parted severely and precisely down the middle. He seldom wore a hat despite the powerful sun that beat down so mercilessly in this part of Africa, and his only acknowledgement of this was a pair of dark rimmed sunglasses that he wore all the time. Gold-rimmed glasses, which hugged his hook-like nose and helped to conceal his penetrating, pale blue eyes that lurked menacingly behind them and gave him the look of a predatory bird.

Despite his in bred killer instinct, Joubert ignored the whine of the mosquito as it hovered over him; he even smiled as the tiny vampire hungrily settled on his hairy forearm and injected its nasal needle in his flesh. Instead of swatting the creature, he pulled the skin on his arm tight so that the mosquito was now unable to withdraw its bloody proboscis. He watched with amusement as the creature teetered on its long spindly legs and squirmed as it tried desperately to escape. Finally in an uncharacteristic act of mercy he released the pressure of his finger and allowed the mosquito to withdraw its probe from his flesh and escape from his arm untouched. Joubert smiled as it rose in the air and flew from the room. It was one of the few times he had allowed an offending creature, animal or human, to escape unharmed, but he had a special affinity for this particular species.

Joubert was an expert on mosquitoes and had immediately recognized the white and dark areas on its wings, which identified it as one of the Anopheles species. Although he knew it to be a carrier of malaria in many other parts of Africa, he was not overly concerned. Effective counter measures that his country had adopted over many years had halted the spread of malaria, many miles further north, at the Zimbabwe border.

He was very familiar with this particular part of the country for he had been born and raised in this desolate northern part of Transvaal province. His father had been a poor white farmer, scratching out a living on a piece of scrubland close to what had then been the Rhodesian border. Despite the family's abject poverty his father had been able to employ several black families as servants on the farm. In those days even the poorest of the white families could still afford to pay the paltry wages that the even poorer black families were forced to survive on.

From the age of six Joubert had learned to treat all black workers with contempt, a piece of human flesh to be abused. Whenever his strict Papa had beaten him for his laziness at school, Joubert had vented his own spite on the black workers by beating them too. As a teenager, the beatings from his father had grown in intensity and frequency but he had learned to tolerate them. Once his beating had ended Joubert would then derive his personal pleasure from giving a much worse beating to a servant of his own choosing…be it man, woman or child. He had graduated to the point that he could thrash even the strongest of his father's workers for he knew that they were too afraid to retaliate.

At that time the power held by the whites over their black countrymen was absolute and for a colored man to fight back or show even the slightest disrespect to a member of the then ruling white race would have meant a more severe beating…or even worse.

Even when his father had finally grown too old and feeble to beat his only son, Joubert had still continued his own sadistic punishment of all the black farm workers now at his disposal. It had now become part of his nature and a thrill that warped his formative years and beyond. Finally upon the death of his Father, Joubert had abandoned his Mother and the farm and had run away to Johannesburg. There he had been able to parley the scholastic achievements that his father had beaten from him into a place at college.

It was here at the Afrikaans speaking University of Witwatersrand that his determination and ambitious thirst for knowledge had served him well. His father had been a harsh taskmaster who had taught him cruelty and sadism, but he had also instilled in Joubert a burning ambition to be the very best at whatever he did.

Commandant Joubert had also served with distinction as the leader of a paratroop battalion in the South African Defense Force, where he had fought in Angola and Mozambique. The capture and subsequent torture of terrorists had been something that both he and his men had excelled at.

However it was a sadistic pleasure Joubert had been forced to curb lately, especially now that President Mandela and his newly liberated comrades had come to power. Men like Joubert had lost much of their privilege and power in the new post -apartheid South Africa.

'Bloody Americans', Joubert snarled, 'Why can't they keep their damn interfering noses out of my country'.

Even though he was secretly opposed to the new regime in South Africa, he still remained fiercely patriotic to the country of his birth.

In fact Joubert regarded the whole continent of Africa as rightly his, both by virtue of his birthright and the ingrained knowledge that his Afrikaner ancestors had fought and died to colonize and develop this once poor country.

Prior to reading his newspaper Joubert had been in an affable mood as he sat on the broad sunny verandah of his spacious house, set amid three lush acres in the wealthy northern suburb of Bryanston.

He had been balancing a plate full of fresh fruit on his lap and the choice ripe mangos and bright yellow cling peaches that he always picked from his garden every morning still glistened with early morning dew. As he selected a succulent piece of mango from the heaped fruit platter he had to restrain his powerful grip so that the rich juice from the soft pulp did not escape through his fingers.

Suddenly discarding the half eaten platter he turned his attention to the swimming pool that shimmered below his verandah, determined to continue with his established routine. The newspaper article had disturbed Joubert's daily ritual and he scowled as he contemplated the consequences of a team of American doctors poking their noses in his corner of the world.

He was not concerned at the pain or suffering that a new Ebola outbreak might bring to a village located deep in the Congo. His concern was that any foreign action or activity that brought unwanted attention so close to his domain could jeopardize his top-secret project, the one he had been working on so diligently for the past two years.

Joubert felt threatened. He had spent too much time and too much effort on his pet project to now allow anyone or anything to stand in his way.

A plan that would ultimately bring his own white race back to its rightful place of dominance, over this vast and potentially wealthy continent.

Only one other person, apparently a high-ranking member of the current government, was aware of his plan, and although this man had never revealed his true identity to Joubert, he had been a regular and very dependable source of money. Covert funding that Commandant Joubert had needed to sustain his secret project over the past two years.

His mystery benefactor was a cunning man, one who like Joubert only pretended to support the new multi-racial government that they were both supposed to be a part of. In fact Joubert's benefactor was only interested in preserving his own position of power and like Joubert, he harbored the same radical views regarding white supremacy.

For this reason he had continued to support Joubert's experiments, experiments that were nearing their completion. Very soon all the injustices of the past few years would be totally and radically put right and his white brethren would once again rule. Joubert smiled to himself at these thoughts.

Shaking off his foul mood, Joubert moved to the little wooden cabana at the edge of his swimming pool and opened the door to check the filtration and pumping equipment inside. This was another one of his morning rituals. Joubert would not begin his self-imposed thirty-lap swim until he had assured himself that the chlorine concentration in the pool was correct and that the filters and pumps were all functioning properly.

As he leaned forward to check the chlorine level on a gauge located just below eye level, he stumbled briefly against the base of the sand filter. As he stopped himself from falling by grasping onto the door, he suddenly spotted the dull red glow.

The red glow was coming from a warning light that had been located innocuously beneath the chlorine indicator.

His body stiffened and his mind raced as he tried desperately to recall when he had last checked this particular warning light. Joubert cursed himself; it was not like him to be so careless.

When he had first installed the light, two years previously, he had checked it each and every morning before his routine swim, but the light had been inactive for so many months and the false sense of security had made him forget all about it. Now its dull red glow brought his mind back to full alert.

The glaring red orb could only mean one thing: some unauthorized person had opened his refrigerator…the special one that was situated in his private laboratory at the Institute.

Only Joubert had access to this particular refrigerator. Even though it was never locked – he had not wanted to give anyone the notion that it contained anything other than routine blood samples – Joubert knew that all his staff lived in fear of him and they were all under strict orders that this particular refrigerator was not to be touched by anyone but him.

Joubert had only installed the devious warning system as an extra precaution and at the time had smiled at his own cleverness. He had been able to install the device so that whenever the refrigerator lid was opened, the slight change in temperature would activate the warning light he had installed in his cabana.

More importantly the electrical signal that triggered this particular light could only be cancelled by pressing a small toggle switch, a switch so cunningly hidden beneath the heavy lid, that only Joubert could possibly know of its whereabouts.

Joubert's anger started to rise now as he reflected on his own carelessness and his eyes narrowed even further as he considered carefully which of his staff might be to blame.

There was only one person in this tightly controlled area that would be stupid enough to disobey his strict orders, the Institute's idle handyman, Sam Nthala.

Nthala was a man Joubert personally despised but someone that he had been forced to employ as part of the new regimes order. The director of the Institute had insisted that even Joubert had to maintain the appearance of supporting equal opportunity.

Joubert dressed quickly. He had now abandoned his early morning swim and this sudden change in his routine made the dark anger fester and bubble up inside him.

As he selected a somber gray suit from his wardrobe his hand momentarily caressed the immaculately pressed blue uniform that hung almost at attention on its hanger. Joubert seldom wore it now but it still had pride of place in his wardrobe. The three colorful rows of ribbons on the chest announced to the world his prowess on the battlefields of Angola and marked all the campaigns he had led in Mozambique. Sadly there was no recognition of his most significant achievement…his membership in the Bureau of State Security.

Joubert had been recruited into the clandestine and notorious organization known by its acronym as BOSS, at the very height of his military career and even though it had to remain a secret, his membership now infused him with a feeling of omnipotence. He was indeed now part of the brotherhood of Afrikaner power.

It was at one of the BOSS secret meetings that he had first met the powerful politician who served as his mentor and to whom he now reported. As he dressed quickly Joubert reflected on all the useful skills he had learned from this feared organization…wire-tapping and demolition techniques were just two of them.

Joubert also had access to a worldwide network of undercover agents that he could call on whenever necessary. However it was the development of his expertise in torture and the extraction of information, both by chemical and physical means, that Joubert had found most helpful: the very special skills that he was going to need later today.

This was something he had always enjoyed and this last thought warmed his mood considerably as he finished dressing and headed for the sleek black Mercedes parked in his driveway. The powerful engine sprang into life as soon as he hit the starter and the fat Michelin tires scrunched on the gravel as he headed for the main road. The very high security gates, which barred entrance to his property, swung open automatically as the car approached and he swept between the wrought iron gateposts.

Joubert leaned forward and pressed a small recessed button beneath the ashtray. Immediately a small drawer popped open to reveal a 9mm automatic pistol. He drew the weapon from its secret drawer and slid it into the inside breast pocket of his jacket where it nestled reassuringly. He had no permit for this weapon and it had no serial number or other distinguishing marks. B.O.S.S. did not concern itself with such trivialities.

Acquiring a weapon, any weapon, had become easy in South Africa, especially now that the country had its own arms manufacturing industry.

Joubert smiled as he reflected on what had happened since the United Nations had tried to impose a ban on the import of weapons into South Africa.

Many new gun manufacturers had immediately sprung up in the Johannesburg area and some had been of such high quality that they had gone on to create a new and thriving export business. The embargo that the rest of the world had tried to impose on his country had been totally ineffectual. An embargo that had only been removed following the release of Mandela and the free elections that followed had actually increased the number of illegal weapons that were now freely available. To get a pistol or even an automatic rifle without bothering to wait out the lengthy process of acquiring an official gun permit from the police department had become an easy task for anyone.

One merely had to bribe one of the poorly paid black laborers who worked in a gun factory, to smuggle out a weapon of choice before it was stamped with its tell tale serial number.

Joubert did not even bother doing this. His many contacts at B.O.S.S. got him any weapon or explosive that he needed, without the formality of any paperwork or gun permit.

In fact collecting instruments of destruction had become a hobby for Joubert and he now had a vast array of weapons. Some of them were stored in his villa; others were kept hidden in his laboratory. Joubert smiled knowingly to himself as his vehicle turned on to the concrete highway and accelerated smoothly in the direction of the Institute of Medical Research. To the site where his secret weapons were hidden and his innocent looking laboratory was located, so peacefully in the leafy eastern suburbs of Johannesburg.

Chapter 2:

The Institute of Medical Research

Commandant Joubert's limousine approached the iron gates of the prestigious South African Institute of Medical Research and the shabbily dressed black security guard manning the area saluted sloppily. Joubert scowled as he reflected on how one of his own hand picked men would have responded to his arrival. His white guards, always dressed in immaculately pressed uniforms, would have sprung to attention whenever he or his official car had approached, but this was the new and liberated South Africa.

However, it served Joubert's purpose that the outward appearance of the Institute was low key and above suspicion. He wrapped himself in this warm thought as he drove to the rear of the compound, which was surrounded by a high razor wire fence. Even this apparent attempt at security attracted little attention for like all buildings in this dilapidated area; it merely looked like a sensible effort to deter thieves. Many people in the surrounding area were so poor that they would have broken in to the compound and stolen whatever they could have laid their hands on.

Joubert screeched to a halt at his reserved space nearest the front door of a squat gray concrete building that was marked clearly with a large letter V. Joubert's laboratory was situated at the rear of the main compound and was in the same location as all the other virology laboratories. To the front of the compound was block A, which housed the animals that were used in various research projects. Guinea pigs and small primates were in constant use at the Institute, for the ongoing study of Tuberculosis, Malaria and Hepatitis.

Block B was much larger and housed all the many departments of Bacteriology that carried out the day to day testing of samples. Samples of blood, sputum and faeces that were regularly submitted from both the local hospitals and from regional Public Health stations scattered across the province of Transvaal.

The alphabetically marked buildings were typical of the style with which government institutions of this kind identified them and the letter on each building denoted their primary purpose by its appropriate letter.

It suited Joubert's purpose that his department was located at the rear of the compound, and because of the high containment viral studies that were conducted in his sector; it also had the highest level of security.

As he strode through the outer door of his block, one of his own men, a highly polished side arm menacingly at his right hip, sprang smartly to attention and saluted. Now Commandant Joubert was within his own empire.

Moving on into the building Joubert passed through a dimly lit corridor that led past the various virology laboratories where technicians scurried about their daily business. He ignored them and continued on until he passed a second guard situated almost at the back of the building. This man, who was called Hans, was a tall and powerfully built ex-soldier. He had served with Joubert as his platoon sergeant in many past campaigns and was unswerving in his loyalty to his Commandant.

Hans too came to attention and saluted smartly, but he also uttered a word of recognition to his superior officer in a guttural but deferential manner. Joubert nodded in recognition as Hans held open the big steel doors for him to pass to the inner sanctum of Block V. This sector currently housed Joubert's own private office and the laboratories under his direct command. The only other person allowed in this section of the building was the general handyman, Sam Nthala.

He would not be here yet, Joubert suddenly realized, for even though this section was supposed to be cleaned before mid morning, Sam was always late. Joubert cursed the fact that he was forced to employ any of these people. Even the menial tasks they were allotted, were done superficially and under protest.

Following the abolition of apartheid there had been a government directive to all employers in South Africa that at least half of all workers hired had to be black. The memoranda that had followed this directive, from the Professor in charge of this state run Institute, left no room for discussion or argument. There was strict enforcement of the new equal employment regulations.

Joubert knew that monitoring of this new policy was also under the watchful eye of the newly appointed assistant director. An ebony skinned man who, in Joubert's opinion, was woefully under qualified for such a high level position. However he was a Xhosa and thus a member of the tribal elite favored by the ruling African National Congress, now in power.

Joubert's mind returned to the large refrigerator once more as he remembered the reason for his early morning visit to his laboratory. He moved quickly to the large white cabinet. Flipping open the heavy lid Joubert pressed the concealed toggle switch, as he had always done before, and gazed over its inner contents.

Deep within the belly of the fridge sat rows of neatly labeled tubes of rich red blood. The stainless steel test tube racks, which each held twelve of the shiny glass tubes, were all lined up with military precision.

Seeing nothing out of place Joubert closed the lid back down again but his brow furrowed as he probed his mind for some other explanation regarding the red warning light. In order to still his nagging doubts he did a tour of inspection of the rest of his department.

Each room was sparsely furnished, as Joubert did not need much in the way of equipment. Any work that he needed done could be delegated to one or more of the other laboratories in the outer reaches of the Institute. Only work that was of a very sensitive or secret nature was reserved for his personal attention and activity. Going from room to room he scrutinized their contents.

One room contained a single piece of equipment, a gene machine he had purchased from a biotechnology company in Israel, two years earlier. This piece of sophisticated technology had saved him a lot of time and effort in the elaboration of the gene sequences on the secret samples he had been working on. A very smart purchase indeed, he thought to himself smugly.

The second room contained a large glass case that he tapped lovingly as he passed by. The vast hoard of mosquitoes within rose up in a dense cloud and whined in protest. These were his new paratroops; these unique flying hypodermics were the cornerstones of his plan.

None of the things in his laboratory would have aroused the slightest suspicion. Even the mosquitoes, which were of the common anopheles variety, were of the type found in many parts of the world.

Although widespread, anopheles mosquitoes were not carriers of the worst disease in Africa…not yet anyway. Joubert smirked as he reflected on this last thought.

Satisfied that nothing was missing or out of place in his secure little sanctum Joubert continued to pace the floor as he awaited the arrival of the idle handyman.

As he waited impatiently he cast his mind back to his youth and gloated on how his special talents had brought him to this place. Following his graduation from the University of Witwatersrand, with a double first class honors degree in Biochemistry and Hematology, he had then gone on to work in the main pathology lab at Johannesburg General Hospital.

He had been a dedicated and brilliant postgraduate student and was soon transferred to the Pretoria Institute of Public Health where he had set up the new department of virology. It was here that he got to meet many high-ranking government officials, somber suited men who trooped in to inspect the new showplace laboratory that was said by many to be the finest in all of Africa.

Joubert worked hard at cultivating these important contacts, always giving them his fawning attention, both during their official daily visits, and at the social events that he personally supervised in the evenings. He was aware of the need to develop all these powerful contacts to ensure that he obtained the funding for the expensive equipment that he needed for his research. With so many political supporters now aligned behind him, Joubert had no trouble justifying the eventual purchase of the electron microscope that he wanted.

This immediately opened up a whole new avenue of research to him. He had long been intrigued by the wonders of genetic sequencing…and its potential for manipulation and modification, now he had the means to probe further into this fascinating new technology. One year later his brilliant thesis on the genetic sequence of the Human Immuno-deficiency Virus had won him the coveted 'South African Gold Medal for Medical Research'. Not surprisingly, it carried with it the offer of a Rhodes scholarship and this additional award offered him the opportunity to study advanced virology at the Harvard Medical School in America.

He was on the fast track now and was all set to embark on this exciting new opportunity. Then he received an anonymous phone call, one night at home. The unknown caller advised him that his future lay with the army of his country, not overseas in America. The mysterious caller went on to say, that if he joined the South African Defense Force he would be given all the funding he would ever need. He could also set up his own private laboratory and carry out research into any area of his own choosing. The caller had gone on to point out that a loyal Afrikaner like Joubert should not accept a scholarship that had been founded by a colonialist like Cecil Rhodes…an enemy of the *volk*.

Joubert had come to realize many years before that even though the 'Anglos' in South Africa, high profile companies like De Beers and the Anglo American Corporation, controlled much of the country's financial wealth, it was Afrikaners who held the political power. Their absolute control over the government, the military and the police ensured and perpetuated this.

For this reason he had not questioned the advice of the mysterious man on the telephone. Instead Joubert had dutifully completed all his tasks at the Public Health Laboratories and then applied for a place in the army.

His fellow workers had been shocked at this sudden career move but they all knew him to be a cold and ruthless man so made no attempts to dissuade him.

One week after his application, he had been instructed to report to the army base at Voortrekkerhoogte. This was the main military base just South of Pretoria. Upon arrival and enlistment there he had immediately been given the rank of Captain. Then following a brief three months of basic training he had been promoted to the rank of Major, supposedly in recognition of his special skills and talents.

Joubert had enjoyed his basic training, especially the unarmed combat; it had given him pleasures he had not enjoyed for a long time. Not since his work at the General Hospital where he had taken a particular delight in the sadistic slaughtering of the laboratory animals, following his test and experiments on them.

The hospital had a strict policy that all laboratory animals were to be disposed of in a humane manner, but when Joubert had been left alone in his own laboratory he had delighted in making them suffer. During the time that he had been in the army he had enjoyed inflicting similar suffering; suffering he imposed on captured terrorists brought in from the border for interrogation.

Joubert's willingness to personally participate in these torture sessions and his delight in this duty instilled in the men under him a sense of both admiration and fear. Joubert was brought back to reality once again by the nagging thought that someone must have violated his orders and opened up the refrigerator in his laboratory.

He returned once more to the squat white case in the corner of the room and this time he peered much more closely at the contents inside.

Certainly there were no tubes missing, but as he lifted out each individual rack of glass vials, to examine them even more closely, he noticed that one tube in particular was different. It had slightly less fluid than the other eleven of its companions in the same rack. This could not possibly be, he thought angrily; only he and he alone dealt with these very special specimens. Ever a stickler for accuracy Joubert always measured out a precise amount of blood sample into each tube.

On closer inspection of the non-conforming glass tube he suddenly let out a loud and spine tingling curse. Not only was the specimen incorrectly filled with blood but it also lacked the identification code that Joubert always inscribed so meticulously on the glass tube.

Uncontrollable rage suddenly now welled up within him as the final realization dawned on him. Someone had indeed violated this forbidden area. They had also somehow stolen one of his precious samples. How could this be, thought Joubert. Only Sam had access to this department and the guard at the door always searched him, whenever he left the building.

Joubert swung round swiftly and strode purposely to his inner office. Once there he unlocked a tall cupboard and took out a dark slim object from within. It was a long black tube, not unlike a billiard cue case. He unscrewed the tube and took out a sjambok, the rhino hide whip favored by old African slave masters. As Joubert coiled the whip expertly around his thick wrist, he sneered. He didn't know how the crime had been committed, but he certainly knew how to find out.

Just then Joubert heard a noise at the outer door and he smiled at the thought that Sam had at last arrived in his sanctum...just in time to help him with his inquiries. As Sam closed the outer door behind him he shuffled sloppily into the room with his mop and bucket dangling loosely on his right arm. The Commandant snarled and moved swiftly to greet him. The first lash of the rhino hide whip cut the skin over Sam's cheek to the bone and as blood started to gush out from his lacerated face, Sam dropped the bucket with a clatter. Sam slid down the wall, blubbering in pain and fear.

'What the hell have you done with the tube of blood that you stole, you stupid fool '? Thundered Joubert. The whip cracked and struck twice more in quick succession. Each strike ripped the thin shirt Sam wore to ribbons. At the same time the thongs of the whip tore apart Sam's flesh so that it now lay open in gushing red welts. Sam's head swam with pain as he slumped like a heap of blood soaked rags on the floor.

The poor handyman dare not fight back for he was in mortal fear of this powerful and angry man now towering over him. Instead, Sam tried desperately to find some excuse -any excuse – for his stupid actions. He had never meant to steal anything; he had been curious as to the contents of the large refrigerator that had constantly tempted his prying mind.

The fact that he was forbidden from opening it only increased his curiosity. Finally he had given in to the temptation.

He had been initially disappointed that it contained only rack upon rack of glass tubes, all containing the same red liquid that he had seen in other parts of the Institute. Still curious, he had lifted out one of the stainless steel racks to pry even further and as he balanced the rack precariously on his arm it had suddenly tilted sideways.

Then before he could prevent it, one of the tubes had slipped out and fallen on to the hard concrete floor. Sam's initial relief that he had not heard it break suddenly turned to panic when he picked it up again. The glass tube was cracked and already the crimson contents were starting to ooze through the thin glass wall. His mind raced as he contemplated the wrath of the man who held absolute power over this area.

Sam quickly seized on a possible solution.

His job as cleaner and general handyman gave him access to all the laboratories in the Institute and he had seen many tubes like this in other refrigerators around the Institute. Sam had no way of knowing that some of their contents were dangerously different; to his mind they were all the same – just tubes of bright red fluid.

It would only be a matter of replacing the broken tube in his hand with one from another area…all he had to do was sneak this one out past the guard at the door. Sam knew that he was always searched when left this particular laboratory…but not when he came back in, so to him, it would only be a dangerous one-way trip.

Sam's suddenly hit on a solution. Quickly dropping his trousers he had quickly inserted the smooth glass tube into his rectum and then, after pulling up his trousers, he had shuffled his way back again to the outer door. Sam had nothing in his hands now except his mop and bucket and the bored sentry on guard at the door had only looked into the bucket before he waved Sam by.

Sam had continued his shuffle through the narrow corridors until he had passed through the outer door of the virology building and crossed the yard to block B.

The security there was much less stringent and after making sure that no one was watching, Sam had quickly dropped his trousers and removed the tube. The foreign object that had been inside him was small and smooth but by the time Sam had reached the door of block B, he was already feeling a stinging sensation in his rectum. Relieved to be rid of the irritation at last he had cleaned the glass tube on a rag from his pocket.

He had then exchanged the damaged tube in his hand for a similar looking tube in a refrigerator nearby. After that it had been a simple matter of returning the newly acquired tube back into Joubert's laboratory.

The guard manning the door of Joubert's laboratory had not even questioned his return; as Sam often had to repeat his cleaning in order to do the job properly. Sam had felt happy once he had put the substitute tube back in Joubert's refrigerator. To Sam's mind all was now as it was before and no one would ever know.

Now Sam's head swam with pain from the sting of Joubert's sjambok and he desperately searched his tortured brain for an excuse to satisfy this fierce man.

The Commandant, sensing that he had not yet inflicted enough pain to extract all the information he wanted, adjusted his grip on the handle of the evil sjambok.

This time he expertly flicked his instrument of torture so that the tail of the whip now remained coiled tightly around Sam's neck. Joubert then deftly flicked the middle of the whip so that it looped around a large hook anchored in the wall. With a powerful heave Joubert hoisted Sam so that he hung like a broken doll.

Joubert eyed Sam's hands dangling helplessly at his side. Noting the large gnarled knuckles on Sam's scrawny right hand, Joubert reached for a set of well-used thumbscrews that he always kept in his pocket. The 'commander of pain' skillfully and expertly applied the dreadful device to the thickest of Sam's dark arthritic knuckles, and tightened the brass thread. The already diseased knuckle made an ominous cracking noise as Sam experienced a pain that he had never felt before. As the screw tightened further Sam's knuckle finally split and the synovial fluid burst forth. Sam was in such agony that his delirious voice now screeched out all the details of his earlier stupidity.

Joubert sensing that he now had all the information that this man had to give drew out his automatic weapon and pressed its snub nose against Sam's temple. Sam felt the cold barrel against his head and screamed as he heard the click of the firing mechanism. This was Sam's last human act as the gun cracked and his poor head exploded onto the wall in a gush of blood, broken bone and white matter.

Sam was now beyond any feeling, including that of the raging infection now spreading through his body.

Chapter 3:

The Virus Crosses the Atlantic

As Joubert strode out of his laboratory, past his white guard at the door, he issued a swift order to his man, Hans, 'There is a mess inside. Get rid of it,' he barked.

Heading off in the direction of block B, Joubert crossed the walkway, and on arriving at the building, kicked open the door with his boot. Sneering at the pathetic lack of security inside he moved quickly past the shabby white-coated technicians who were too busy with their routine business to even notice him.

Striding to a large refrigerator at the back of the room, the one that Sam's tortured lips had moments before directed him to, he flung open the heavy insulated lid. Just then one of the technicians, who had been working at a bench nearby, noticed Joubert and called to him.

'If you are looking for the sample that was due to be shipped out yesterday Sir...It went out this morning...by priority post.' The technician had recognized Joubert as a man of some authority and as he stood there, almost at attention, he awaited a response.

Joubert looked at him questioningly, his raised eyebrows inviting the man to continue.

'Yes Sir, one of the vials was cracked and leaking, so I had to combine both samples and then transfer the fluid that was left into two new tubes. This caused some delay, so I put a rush on the order...especially as it was going overseas'. The technician smiled now, proud of the fact that he had used his initiative in fixing a problem, and got the shipment out on time. Joubert realized immediately what had happened. The vital specimen he now sought had somehow got mixed in with samples that the Institute often shipped to research labs and teaching Institutes around the world. These items were a valuable source of revenue for the Institute but Joubert's only concern now was in tracing the whereabouts of his missing offspring, and either its rapid recovery or complete and utter destruction. Joubert somehow maintained his calm composure as he casually asked the technician if he could recall the address where the shipment in question had gone.

18

The technician apologized; he did not recall the address, but only that it was somewhere in America. Joubert thanked the man curtly, swung on his heels and headed back to his office. He had all the information he needed to put some type of recovery plan into place.

Reaching his own laboratory once more Joubert passed his guard at the outer doorway and the man uttered one sentence. 'All is secure and in order, Commandant.' Joubert did not have to look in the direction of where Sam's limp body had dangled dead on the wall, only thirty minutes before; he knew that his efficient subordinate had removed the body and all traces of it.

Joubert also knew that very few questions would be asked regarding the whereabouts of the missing black handyman. It would just be assumed that he had merely returned to his home and family in Swaziland.

Joubert entered his office and flicked open a drawer. It contained a keyboard that connected him directly to the main computer that served all departments at the Institute. With a few deft keystrokes he called up a master program that detailed all shipments that had been sent out of the Institute over the past 24 hours. There was only one bound for the United States and he jotted down the name of the purchaser and the address of the hospital in Detroit, where it had been shipped. He then placed a direct phone call to the South African embassy in Washington.

On hearing the familiar sounding message at the main switchboard, Joubert immediately pressed a four-digit extension number he had memorized months before. He was rewarded with the familiar sounding voice.

Captain Devere, an obedient ex soldier who had served under Joubert in Mozambique had just recently been recruited by BOSS He was now one of their many undercover agents in America. Joubert issued a few brief instructions and then gave his agent the name and address from the pad in front of him of the man who had ordered the package and his address in Detroit.

Joubert replaced the telephone receiver. All that could be done had now been done and as Joubert leaned back in his chair he felt confident that he could rely on Captain Devere to either recover the missing tube or to destroy its contents, exactly as he had instructed him.

With this matter now in hand Joubert rose from his desk and turned his attention to the familiar feelings that were welling up in his loins. He had enjoyed the torture and eventual destruction of Sam but it had all been over too quickly. Better men would have lasted hours before confessing, Joubert sneered to himself.

The emotions that he had experienced were still swirling in his head and he realized that they would have to be satiated before getting back to his routine.

Grabbing the keys to his car, he strode out to the parking area and was once more seated in the plush leather interior of his sleek automobile. As he urged the car through the gates of the compound he did not spare a thought for the problems that were behind him. He was now being driven by very different desires.

As his car reached the motorway he was soon moving swiftly though the mid morning traffic that was now headed out alongside him to the Northern suburbs. Unlike many of the luxury cars that traveled beside him, he ignored the familiar turn off into Bryanston and continued on instead around the major highway.

As he started to smell the familiar wood smoke of the shanty-town that was now spread out on his left hand side he closed the electric widows tight and looked with disdain at the ramshackle buildings ahead of him. He knew that he was fast approaching the outer limits of the sprawling township that was known as Soweto. He pressed the electric window control again to make sure that the windows were shut, to keep out the smell.

At the far reaches of the huge township he turned right on to an unmarked dusty track and then headed straight for a large clapboard house that stood apart from all the other smaller shacks. On reaching the driveway he brought his car to a halt in a cloud of brown dust. Jumping out from his car Joubert gazed longingly at the crude hand painted shingle that hung on the wooden door. The sign simply announced that this was the home of Madame Lazelia but it gave no details of her services or the type of business conducted here.

Joubert had only been here twice before, but on both occasions it had been at night. His previous nocturnal visits had left him somewhat unprepared for the garishness of this purple and white building that now stood before him shabbily, in the very bright sunlight. He was now forced to note that the scarlet curtains, which on previous visits at night had given out a warm inviting glow, now appeared shabby and uninviting. But Joubert was not concerned with the buildings outer appearance. He only knew what pleasures were waiting for him inside and he bit his lip in anticipation He ran up the stone stairs and rapped on the door.

The door opened almost immediately to reveal an immense black woman who stood over six feet tall. Her jet-black hair was piled high upon her head in an old-fashioned beehive style and this added even more height to her already tall and commanding stature. She wore a loose multi-colored kimono over her obviously well endowed body and as she smiled, gold fillings scattered in her white teeth, glistened in the sunlight.

'Welcome, Sir, It is so very good to see you again'. The woman showed no surprise at his unannounced visit and as she closed the door behind him she led the way into a dimly lit parlor where the smell of burning incense filled the room with its suffocating sweetness. Motioning him to a mock leather armchair Madame Lazelia poured a glass of red wine from a half-empty bottle standing on a table in the middle of the room.

Joubert took the glass of wine and drained it in a single gulp but instead of sitting down in the chair he went over to a large circular bed in the corner of the room and started to undress.

'I see you are anxious to begin,' she drawled, and as he continued to disrobe Lazelia disappeared behind a set of beaded curtains at the other end of the room. When she reappeared her kimono had been discarded and she had loosened her hair so that it hung like black drapes that swirled around her large pendulous breasts.

With the kimono now gone, she was dressed only in a black leather apron that was affixed around her broad hips by narrow thongs, each one tied in a neat red bow. Her black thighs glistened in the dimly lit room and as she tottered back toward him Joubert noted how the muscles in her calves bulged invitingly and were emphasized by the high-heeled stiletto shoes she had obviously just put on. Joubert had undressed to a pair of khaki boxer shorts and the rest of his clothes had been discarded in a crumpled heap on the floor.

Without speaking he climbed onto the bed and lay face down on the bright scarlet duvet. Madame Lazelia also climbed on the bed beside him and wrapped her powerful thighs around his hips. She then held him in a vice-like grip with her legs, as she began to massage his back. Moments later her fingers started to probe deeper into his skin.

As the sensation he was now feeling became even more intense, her long sharp fingernails started to gouge deep down into his flesh. Small trickles of Joubert's blood followed each stroke of her sharp nails and as she leaned even further forward, to gain more leverage, her huge breasts with their now erect nipples, brushed across his broad back.

Joubert groaned with pleasure and she responded by reaching above her head to unhook a small leather flail that was hanging on the wall.

She began to beat him with it so that the brass fixtures at the end of each thong grazed his flesh initially, but as the whiplashes grew in intensity they cut into his skin and left large angry purple bruises.

Joubert squirmed with delight and rolled his body in a conflict of agony and ecstasy. Finally when he was beginning think he could stand the pain no longer she pulled down his boxer shorts and lashed ferociously on his previously untouched buttocks. Joubert groaned and heaved a shudder of intense pleasure as his swollen phallus jerked out its own final satisfaction.

Lazelia climbed off the bed and picked up the bundle of money that he had left for her on the table. As she disappeared back behind the beaded curtains Joubert still lay there on the bed, reflecting on the heady excitement of the day.

Chapter 4:

Detroit, Michigan

The Detroit Municipal Hospital was an unimposing, gray brick building situated in the downtown core of the city. Surrounding the hospital and its two small car parks, a mess of high-rise apartments towered over the drab hospital. Many of their occupants overlooked the laboratory on the fourth floor where Bill Williams had toiled loyally, for the past twenty-two years.

Bill had begun his career here as a student technician and over the years had risen through the ranks to the point that he was now in charge of the laboratory.

In addition to his ever-increasing administrative duties he still performed routine work in the department of microbiology, which was a way of keeping up his technical skills and was an area he found interesting.

Bill was a plump, cheerful figure, well liked by all his staff and who, because of his years of experience, was often called on to share his knowledge with all the more junior members of his department. He particularly enjoyed the teaching aspects of his job, for in addition to running seminars within the hospital; he also conducted workshops at many of the annual interstate meetings, which was a welcome break from his grinding routine.

The budgets had been very tight lately so he had been relieved when his boss had given approval for him to attend and present his parasitology seminar at the upcoming American Biology meeting that was scheduled to begin the following day in Washington.

As he busied himself around the lab, clearing up the last of the routine samples that he had processed that afternoon, he glanced anxiously at a clock on the wall. Bill realized that he did not have a lot of time left if he was to catch the last flight out of Detroit tonight.

Bill had packed his overnight bags earlier and they were already stored downstairs in his car, but he still had to prepare the samples for his workshop the next morning. He liked the samples for his eager students to be fresh so had left the task of preparing them until the last moment.

With the last of his patient reports finally finished he swung open the door of the big refrigerator and took out the package that had arrived from South Africa that morning. Placing the package on the bench in front of him he noted that the wrapper was marked 'Priority Post' and it also had a bright orange label that screamed out 'Handle with Care, Biological Sample, for Teaching Purposes Only.'

As he snipped open the outer wrapping with a pair of sharp surgical scissors, he thoughtfully placed the bright stamps from the parcel on a bench nearby, they would be much appreciated by his nephew who was an avid stamp collector. Bill extracted the two tubes of blood from their protective packaging and carefully placed them in a stainless steel rack, alongside another rack, that already contained twelve clean test tubes.

He then checked the workshop enrolment sheet on the desk in front of him to confirm that he did indeed have twelve registrants for his workshop, the following day. It would be nice to have one specimen for each of his students, thought Bill; they could individually process the wet blood preparations and see for themselves the malarial parasites that he knew each of these twelve new samples would soon contain.

In the past he would have run a quick microscopic check on the two master tubes but he omitted this step today as he was rushed for time. He had no reason to doubt the quality and contents of these specimens, he had ordered similar ones many times before, from the South African Institute in Johannesburg, and they had always been absolutely dependable.

Donning a pair of rubber gloves he used a long glass Pasteur pipette to suck up an equal quantity of blood from each of the two larger tubes in turn. This sample he then dribbled slowly into each of the twelve smaller tubes so that they all contained equal amounts of fluid.

He was just about to spin down the twelve portions of blood, in preparation for their journey to Washington, when there was an urgent knocking at the outer door. Bill looked up from his task and saw the broad smiling face of Rick Harvey poking round the door.

Damn it, thought Bill, he had completely forgotten his appointment with Rick that afternoon, then he remembered that this particular salesman had only dropped by to pick up a purchase order that Bill had promised him. Bill relaxed a little; it should only take him a few minutes to deal with this minor interruption.

Bill had agreed to purchase a new hematology analyzer from Rick, two weeks earlier, and he had agreed to have the order ready for Rick this afternoon, so that the instrument could be delivered the following week. 'Please take a seat' said Bill, stripping off his gloves. 'I'll just nip through to my other office and get you the purchase order Rick. It is all ready and waiting'.

Rick smiled and sat himself down on the same stool where Bill had just moments before been working.

The young salesman grinned at the thought of yet another instrument order in the bag and one less to meet his quota.

As Bill left the room and headed for his outer office, Rick idly cast his eyes over the bench and his gaze suddenly fell on the twelve little tubes that were sitting in their secure little stainless steel holder. The sight of them jogged Rick's memory as he remembered that he had arranged to do an instrument presentation at the nearby Henry Ford hospital, the following Monday.

In his haste to get away for the weekend Rick had neglected to call in at his office and pick up some test material for this very important demonstration, Simulated blood he would need, if he were to impress the largest and most influential hospital in Detroit. Rick silently cursed his own forgetfulness, it was too late now to drive back to the office, especially if he wanted to miss the heavy rush hour traffic in Detroit and be well over the border into Canada before dusk.

Rick had been looking forward to spending the long weekend at his friend's cottage in Muskoka, a small country town, just north of Toronto. He had purposely planned his last call on Bill so that he could pick up the purchase order he needed and then be well on his way before rush hour. The traffic built up quickly in Detroit and he wanted to be over the Canadian border and beyond the busy city of Toronto, before nightfall.

Almost without thinking Rick leaned forward and picked one of the glass tubes from the rack and slipped it into the top breast pocket of his blazer jacket.

At that precise moment it never occurred to him that the little tube might be missed, or that he was stealing. As Rick sat back once again on his stool, his foolish action of the moment was immediately forgotten.

Moments later Bill returned to the laboratory clutching the purchase order and Rick rose from the stool to grab it. As soon as he had stuffed it into his briefcase Rick gave a quick nod of gratitude and then sped out the door. At last he was done for this week with his sales calls, thought Rick, now he could head out of the city and be on his way to the welcome weekend break that he had been promising himself.

Meanwhile Bill had returned to the task that had been occupying him before he had been interrupted, but he looked puzzled as he spotted the vacant space in the stainless steel rack. It never occurred to him that one of his samples, might have been taken. These samples have no real value thought Bill to himself; I must have miscounted them in my haste to get the job finished. Bill cursed under his breath; he would now have to reallocate blood from the eleven remaining tubes in order to make up for the apparent discrepancy, before spinning them down for their journey to Washington.

He glanced up at the clock again and realized that time was now fast running out before the departure of his plane, especially if he encountered any kind of delay on his way to the airport. Flustered now, he selected another clean pipette and hurriedly transferred a little of the blood from each of the eleven remaining tubes into a fresh clean tube. Once the new tube contained the same amount of blood as the other eleven, he placed the twelve blood specimens into the centrifuge and flicked on the power switch of the machine.

As he closed the lid the Medicalert bracelet that hung loosely from his wrist clanked against the metal cover. It was a constant reminder to Bill and to the world that he had been diagnosed as a hemophiliac.

Bill had been diagnosed with this disorder as a child, but it was only a minor inconvenience to him now, especially since he had been getting his regular injections of 'Factor Eight concentrate' at the hospital clinic nearby. This vital therapy replaced the clotting factor that his blood had sadly lacked from birth, and prevented the dreadful bleeding episodes he had suffered as a small child.

The squat device before him started to hum as it picked up speed but then as Bill was turning away to gather up his notes, there was an ominous clunk and loud clatter. The whispering whine that the machine had been making moments earlier had now changed to a dreadful clatter. It could only mean one thing, thought Bill, as he quickly turned and turned off the power…a tube had broken inside the rapidly rotating centrifuge.

Bill cursed out loud this time; the additional delay that this would bring to his already very tight schedule was now making him angry and flustered. Even before the head of the centrifuge had stopped spinning completely he reached into the machine and started to extract the renegade tube that had shattered inside its stainless steel cup.

As he hastened to repeat the very process he had just completed moments earlier, he felt a jab of pain as the sharp edge of the broken tube pierced his finger. At that very same instant Bill realized that he was no longer wearing his protective rubber gloves.

Thank heavens, I had my factor eight shot last week, thought Bill, he knew that his finger would not bleed excessively but he would now have to sterilize both his finger and the whole damn centrifuge, before he could even leave the laboratory and head out to the airport.

Bill ran to the corner of the room and grabbed a large bottle of bleach. After first removing the rest of the unbroken tubes from the now silent centrifuge, he swabbed both the inside of the machine and the surrounding bench.

This all took several precious minutes before he could turn his attention to his lacerated and still bleeding finger. Bill clucked with impatience as he splashed the harsh bleach on his finger and then put his hand under the cold-water tap, just as the stinging sensation of the bleach hit his brain.

As the pain quickly subsided under the cold jet of water, he dried his hand on a paper towel and then quickly applied a band-aid. Gathering up all the remaining samples and placing them inside a plastic container, he put them all into his briefcase. The plastic travel container was made of light white Styrofoam and it would keep his blood samples cool and secure during their short journey to Washington.

Two students would just have to share one sample, he rationalized as he stuffed all his lecture notes hurriedly into his briefcase and at last headed out of the door. Bill's injured finger throbbed only slightly as he rode the elevator down to the ground floor and as he strode briskly to his waiting car…his attention was now focused solely on beating the already mounting traffic.

As he clicked on his seat belt his swollen finger caught on the buckle and caused him to wince in pain, but Bill was too intent on reaching his destination to notice.

He was also in far too much of a hurry to reflect on the fact that some infectious agent from the broken tube might already be invading his bloodstream.

As he accelerated rapidly out of the hospital driveway Bill did not spare a passing glance at the grim faced man in the black BMW, who almost ran into Bill as he headed so purposefully to the visitors' car park nearby.

Chapter 5:

First Explosion

The man in the black BMW was angry and frustrated as he pulled into one of the many vacant spaces that were now available in the visitor's car park. He had left Washington in plenty of time but the traffic on the freeway had been slowed by an accident involving a tractor-trailer that had somersaulted over the embankment and spilled its load across all three lanes of the usually fast moving highway.

This unexpected delay had prevented Captain Carel Devere from reaching this insignificant Detroit hospital by mid afternoon and had screwed up his plans to reach the man he urgently sought today.

Now it was late and his quarry might have left the location that Devere had been given. As he swung his legs from the car he reached under the seat and took out a small white bundle concealed there. As the garment unrolled on his lap Devere took out a silver tube that was cushioned in the middle of the hospital uniform and stuck it firmly in the belt around his waist. He continued to unroll the white garment and donned it over his street clothes. Then from the pocket of the laboratory coat that he was now wearing, he extracted a stethoscope and clipped it around his neck so that it dangled on his chest like a badge of office.

As he moved toward the hospital entrance he knew that this simple disguise would give him unchallenged access to all parts of the hospital, especially now that most of the day time staff had already gone home.

Devere hardly missed a step as the automatic doors slid open silently, and as he passed by the notice board in the main lobby he glanced up to confirm that the department he sought was on the fourth floor. None of the staff he passed gave him a second glance but he ignored the elevator that was already in use and ducked through a doorway that led to the stairs.

He took all the stairs; two at a time, and by the time he reached the fourth floor he still breathed effortlessly and his pulse remained steady. Passing swiftly down the gray walled corridor he spotted the door marked 'Bill Williams, Chief Technologist' and without knocking he entered the small office that led into this area and surveyed the room within.

There was nothing much of interest to him here so he moved on into the laboratory that Bill Willams had left only minutes before. His eagle eyes spotted the familiar South African stamps that were still lying on the bench and he pounced on this tell tale clue to the item he sought. Grabbing the package nearby Devere crumpled it up with disgust as soon as he realized that the contents it had once contained were now gone.

He continued to search the whole area quickly and professionally but apart from two empty glass tubes and a centrifuge that smelled strongly of bleach, there was no trace of the prize he had come so far to find.

Damn it, thought Devere, the man I just missed must have taken it with him and as he moved back into the outer office he clenched his fists in suppressed rage.

The supposedly simple task he had been given by his master in Johannesburg was now making him angry.

In order to determine the next possible whereabouts of his quarry he was about to root through the papers on the desk when his eyes eagerly fell on a copy of an airline invoice. It was neatly clipped to a programme that gave details of a meeting that the man he sought must be attending, at the Hilton Hotel in Washington.

Devere's eyes narrowed to slits as he noted that one item on the program had been highlighted in red ink. He smiled with satisfaction as he read the name of Bill Williams, listed as the seminar leader for the parasitology workshop scheduled for the next day. How thoughtful, his prey had certainly left him a nice clear trail to follow, Devere sneered as he pocketed the program and the hotel confirmation slip stapled to it.

As he was leaving Devere took the silver tube from his belt, tore the yellow tab from its base and flung the thing into the laboratory he had just finished searching. The silver canister landed with a metallic thump on the sturdy wooden bench and quickly fizzled into life as its incendiary contents burst forth and filled the room with a blinding white light. Even before Devere had reached the top of the stairwell, the white hot candle had already burnt its way though the solid oak bench.

By the time he reached the bottom of the stairs, the muffled explosion that followed, told him that the small but efficient firebomb had ignited some of the solvents that had obviously been stored nearby.

As Devere reached his car the fire alarms were already clanging angrily throughout the hospital and as he looked up he saw with great satisfaction that the fourth floor above him was already a raging inferno.

Devere had been fairly certain that there was nothing of interest left behind in the laboratory, now he was sure. Even if his brief search had missed anything…anything at all, whatever it was would now be totally destroyed.

Once back inside his car Devere peeled off the now unwanted white coat and flung both it and the stethoscope, on to the back seat of the BMW.

Satisfied with his destruction of the laboratory, Devere accelerated the swift little vehicle out of the hospital car park and headed in the direction of the airport.

He was close on the trail of poor Bill, who meanwhile had no way of knowing that the laboratory where he had labored loyally for the last twenty years, was now just a smoking heap of rubble.

Bill also had no knowledge of the identity of the man who was now relentlessly following him, or the hatred that coursed through the veins of his grim follower.

Chapter 6:

Detroit to Toronto

Rick Harvey was also unaware of the drama that had unfolded behind him. Since leaving the hospital he had happily negotiated the traffic through Detroit and passed without challenge through the American border post. At the Canadian border he had only waited in line for a few minutes, as each car ahead of him edged through the checkpoint, before it was his turn to pass through. As Rick drew up alongside the little glass kiosk the tired border guard glanced at Rick and then scanned the back of his empty car before asking him the standard questions.

'Anything to declare?. Any guns or alcohol?'.

'No nothing at all,' Rick responded politely.

'How long do you plan to remain in Canada and what is the express purpose of your visit?'

'I'm only here for the weekend, taking a short vacation at a cottage in your beautiful Muskoka region, just north of Toronto', Rick replied, smiling now.

The guard did not return the smile, but waved him by with a curt 'Have a nice trip Sir'.

Rick breathed a sigh of relief and drove on through the border post, he had nothing to declare as it happened, but it was always a tense moment passing through the border. Border guards had absolute power and if they were not satisfied with the response to their standard questions they would direct any vehicles that failed their cursory approval to the nearby border station.

Here more officious guards would ask even more searching questions. It was here that the verification of documents and a strong possibility of a car search would mean considerable delay and inconvenience.

As Rick drove out of the border area he passed by a huge sign bearing the words 'Welcome to Ontario' then he spotted another sign pointing in the direction of the Queen Elizabeth Parkway. Turning in the direction indicated, he accelerated down the access ramp and was soon speeding along the fast four-lane motorway heading for Toronto.

As he settled back more comfortably in his seat he wound up the window and switched on the car radio. His blazer containing the stolen tube of blood had already been tossed carelessly onto the back seat.

Now he was fully relaxed and free to enjoy the start of his vacation. Rob sat casually in his shirtsleeves as the car burned up the miles en route to his eventual destination, and all he was thinking about now was his forthcoming weekend break from work and the pleasures of relaxing at the little cottage by the lake.

Whizzing past one of the service areas just off the busy highway, he glanced down at his fuel gauge and made a mental note to stop in at the very next rest area. There he could fill up the gas tank, and grab a quick coffee and sandwich. His speedometer was hovering around the 120-kilometer mark and it didn't take long before he saw the sign announcing the location of the next refreshment area. As he drew closer he eased up on the accelerator and moved into the inside lane.

On reaching the entrance to the combined garage and restaurant, he swerved off the highway and drove in. Parking his car in a vacant spot near the coffee shop Rick switched off the ignition and hauled his jacket from the back seat.

Checking that his wallet was still secure in the inside pocket of his blazer, he failed to notice that the little glass tube that he had stolen from the laboratory had slipped from his pocket, and was now lying on the floor beneath his seat.

Fifteen minutes later, with a warm coffee and a hot chicken sandwich under his belt, he only stopped once more, to fill up the gas tank at the petrol pumps nearby. Then he was back on the fast highway and on his way once again to Toronto. On reaching the outskirts of the big modern city, he could not help noticing the massive concrete structure of the Toronto Tower, outlined on the skyline, announcing itself as the tallest in the world As he negotiated his car around the inner ring road he inwardly blessed the city planners; as this road would avoid the busy downtown core of the city.

Following the signs that directed him to highway 400, which he knew would eventually take him North to cottage country, Rick smiled, he was on the final leg of his trip. Once on Highway 400 he continued to make good time, even though he had now been joined by a stream of other vehicles, which were also headed out of the city in search of a day of rest and relaxation.

The Muskoka area was one of Ontario's most popular tourist destinations as its 1,600 lakes provided ample waters for bathing, boating and fishing, as well as providing the peace and tranquillity that Rick and his fellow city dwellers needed from time to time.

Rick knew that he was nearing his final destination when he noted that the green fields and forests around him were now starting to erupt with pink granite outcrops beside the highway he was now driving on. Crossing the bridge over the Severn River he passed by the Muskoka Tourism Welcome Center but he had no need of maps, this was an area he knew very well.

He was anxious now to reach the calming, clean air of his little cottage before nightfall. Turning at last off Highway 11 he meandered along a small secondary road for several miles until he spotted a familiar lake on his left. Driving round the clear waters edge, which was flanked by tall green trees, he saw that many of the boathouses lining the shore were quite beautiful. Many of their windows were adorned with boxes of brightly colored flowers.

Some of the larger houses alongside the shore were quite elaborate, having been built by very wealthy Americans wanting to escape their crime-ridden cities.

They certainly did not fit into the image of a 'cottage' that Rick had, he reflected to himself, enviously.

Finally he pulled up outside his own wooden cottage, the place where he could relax for the next two days.

Stretching his arms and legs with relief, Rick absent mindedly pulled on the seat mechanism below his knees and moved his seat back to accommodate his long aching legs. The crack of the fragile tube, as it was crushed against the harsh metal frame of the seat, was drowned out by the sound of his blaring car radio.

As Rick stepped out from the car, the red fluid was already starting to ooze from its broken container and as the blood seeped into the carpet, its aroma attracted a pair of hungry mosquitoes that were dancing in the bushes nearby.

With an excited whine they both headed in the direction of this apparent banquet of opportunity and started feasting on the red nectar that was now fast disappearing into the thick pile of the carpet.

Satiated at last, they withdrew their blood filled snouts from the congealing mess on the floor and winged there way back outside through the open car window. Within seconds they had disappeared once more into their woodland hide.

Chapter 7:

Washington, DC

The jet was already circling Dulles airport and Bill Williams was polishing off his second glass of red wine as the heavy landing gear locked into place. Sluggishly he tried to determine if the perspiration beaded on his brow was due to his last minute dash for the plane or to the glasses of Cabernet he had so quickly consumed. He threw off the dull headache and concentrated instead on the evening that still lay ahead of him.

With luck the woman he had met at last years meeting would be there again this year, and if all went well he would not be sleeping alone tonight. She was a happily married woman but Bill did not dwell on that thought too long. Instead he smiled as he recalled that even though she was middle aged, she was still in pretty good shape and she would surely be a welcome diversion this evening.

As bawdy thoughts continued to occupy him, Bill passed through the terminal, grabbed his luggage from the still revolving baggage carousel and hailed a waiting cab outside the airport. The cabby loaded Bill's garment bag into the back of the car but Bill kept his briefcase with him, as he had done on the plane.

The taxi headed downtown and Bill opened up his briefcase, to reassure him that the samples inside were still upright at the bottom of his bag were still wedged securely between his lecture notes. The cab was hot so Bill wound down the window to let the cool night air blow across his fevered brow, but he was still feeling out of sorts as the taxi drew up outside the bright lights of the Hilton hotel.

Paying off the cabby and ignoring the young bell-man hovering nearby he shouldered his canvas garment bag and ambled through the doors that led to the hotel.

At the front desk he resisted the temptation of asking the desk clerk if the lady he was seeking was also staying at this hotel. Better to be discrete and phone her from my room later, thought Bill.

Within a few minutes he was through all the hotel formalities, had the key to his room and was ascending the half full elevator to the ninth floor. He hardly glanced at the other gray-faced road warriors in the elevator who, like him, were intent only on getting to their rooms. Room 967 was located to the right of the elevator and as he unlocked the heavy door he slung his garment bag on to the queen size bed and snapped on the light.

After unpacking his clothes and hanging them in a closet by the bathroom, he returned to his briefcase. Opening it up he took out the package containing the precious tubes of blood and placed it on a table next to the telephone. This done he picked up the phone and after a brief inquiry, slammed the phone back on its cradle with a shrug. There was no record of a Betty Johnson staying at the hotel, the desk clerk insisted.

Maybe I'll get lucky down at the bar, thought Bill. Bill showered quickly and as he shaved in the mirror he grimaced at his reflection. It was now bloated and blotchy. Even after his shower his face looked sallow and there were still beads of perspiration on his cheeks. I must be tired, thought Bill. A cool glass of beer would help and after dressing quickly, he left his room and descended the elevator to the ground floor bar.

The bar room was brightly lit and a lone piano player at the far end of the room banged out a familiar seventies tune.

It did little to cheer the half empty room of weary travelers, who sat sipping their comforting drinks. Bill had only been seated in the bar for a few minutes, a drink already in his hand, when a young girl in her early twenties approached him.

He could not help but notice the low cut and very revealing green dress she was wearing as she glided over to his table and sat down in the chair beside him.

Bill no longer had any delusions regarding his age or his appeal to the opposite sex. With only two glasses of wine and half a beer under his belt he knew right away that this voluptuous creature sitting next to him was a hooker. Although he was not in the habit of paying for sex, he rationalized that it had been a long tiring day, and she was certainly young and attractive.

His head swam with the heady perfume she was wearing as she leaned forward seductively; she placed her hand on his knee and negotiated the $100 it would cost Bill for an hour of her time back in his room.

With no thoughts now for the other patrons around him Bill drained the last of his beer, dropped a ten-dollar bill on the table and left the bar with this fresh young delight hanging on his arm.

They rode the elevator together silently but once inside his room she brazenly held out her hand. 'I want the money first please. Before we start.'

Bill extracted two fifty-dollar bills from his wallet and thrust them into her open hand. As she stuffed them into her purse she spun on her heels and headed for the bathroom but Bill grabbed her arm possessively.

'For that money, I want to undress you myself'.

The young prostitute smiled at him submissively, kicked off her high-heeled shoes and turned around wantonly so that Bill could access the long zipper at the back of her dress.

Bill pulled the zip down slowly so that the flimsy dress slid smoothly from her shoulders and snaked over her hips to the floor. Bill hungrily planted his lips into the soft nape of her neck, at the same time as he reached both his hands forward so that they cupped her full round breasts.

As he continued to caress her warm flesh he felt her nipples harden and rise between his exploring fingers and they become even more erect as his fingers stroked them. She squirmed slightly as his lips moved away from the nape of her neck and licked their way longingly under the curve of her full breasts. He found one nipple with his searching mouth and sucked it vigorously.

After a few moments Bill pushed her away from him so that he could gaze fully at her young naked body. She was only in her early twenties and her figure would have done justice to the centerfold of any *Playboy* magazine. His lustful eyes traveled downward from her out thrust, pink tipped breasts, down to the neatly shaved black triangle below her flat stomach.

Apart from a small blue butterfly, tattooed above her right breast, her skin was flawless and Bill quickly forgot about his throbbing head as another throbbing desire welled up in his loins.

The young girl, sensing his rising passion, undid the belt around his waist and as his pants dropped to the floor she grasped his pulsating manhood with both hands.

Bill groaned now as she expertly manipulated the tip of his penis and he had to tauten the muscles in his thighs to stop himself from ejaculating prematurely. When she began to rub his suffused organ against the soft pubic hair between her legs he could resist no longer and pulled her urgently toward the bed.

He fell back onto the bed and as she landed on top of him, her breasts pressed erotically into his naked chest. Bill was sweating heavily now as he sought to reach down with his hand and guide his erect phallus into the warm dark depths between her thighs. His young partner was not quite as caught up in the passion of the moment and before she would let him penetrate her she reached for the purse that she had strategically placed on the bed.

Quickly and expertly the young hooker took out a condom and rolled it onto him, before he knew it. Allowing him to continue they both writhed together so frantically that neither Bill nor the young hooker heard the click of a plastic credit card being slipped into the lock of room 967.

Moments later, Captain Devere slipped quietly into the dark room only pausing to allow his eyes to become accustomed to the gloom. Satisfied that the two bucking bodies on the bed were otherwise occupied, he moved stealthily to the package that he had already spotted on the table. The vague light from the window had illuminated it eerily and he guessed what it was immediately. In his haste to grab the package his coat sleeve caught on the telephone receiver and as it clattered noisily from its cradle, one of the figures on the bed rose up and peered in his direction.

Swift as a panther, Devere leaped to the bed but the naked woman was almost as quick and rolled to one side as Devere grabbed the jaw of the man who was lying there. Bill William's had been far too occupied to react to this sudden threat and he lay there with his mouth wide open in surprise as Devere held his jaw in a vice like grip.

Bill Williams was unable to utter a sound as Devere drew a hypodermic from his belt with his free hand and plunged the long needle deep into Bill's exposed neck. It was all over in seconds.

Poor helpless Bill didn't even feel the pain as the thrombin in the syringe invaded his flesh and fashioned several large blood clots within his violated bloodstream. As more of the powerful blood-clotting enzyme reached his brain, Bill expired silently, just as Devere had intended.

Devere now turned his attention to the naked woman who was still cowering but as he lunged toward her she scrabbled on the floor and her sweaty body slipped through his outstretched clawing fingers. As he moved around the bed, to corner her against the wall, she sprang across the pillows and reached the door, which was still ajar. Fearing for her life, she ran naked from the room.

Devere composed himself. It was only an insignificant hooker he thought. He now had the package and the man who had led him on such a wild goose chase, the man who was now lying dead on the bed, would trouble him no more.

As he surveyed the scene Devere smiled as he thought of the obvious conclusions the authorities would draw. Some overweight and out of shape businessman had taken a young hooker back to his room and at the very height of his passion had suffered a fatal heart attack. Not unusual, thought Devere. He was also smugly confident that the extract of human thrombin that he had so expertly injected into Bill would be so similar to the dead man's own normal clotting enzyme, that even if an autopsy were performed, its presence would go undetected.

As he looked with disdain at the pathetic figure now lying dead on the bed before him, he had no way of knowing that the swift end that he had meted out to this man had in fact delivered him from a much more painful and lingering death. Just like as his superior, Commandant Joubert had done to a poor handyman half a world away.

Devere glanced around the room to be sure that there were no traces of his brief passing and then gathered up the package he had come so far to get. He left the room as quickly and unobtrusively as he had arrived.

Chapter 8:

The Washington Hilton

The night manager on the desk of the Washington Hilton was surprised but not shocked when a naked young girl fled through the lobby. The hotel tried to discourage hookers from plying their trade in the hotel but it was an ongoing battle. As he shuffled through his boring routine, the young manager shook his head and returned to his work; he had no concern for what hotel guests did in their own room, as long as it didn't cause him any problems.

The minor incident momentarily forgotten, his mundane task was once again interrupted by the jangle of the phone at his elbow. He scowled as he picked up the receiver and listened to the angry voice at the other end of the line. An irate female guest on the ninth floor was angrily articulating her concern regarding a disturbance that had occurred outside her room on the ninth floor.

The night manager sighed. It was strict hotel policy to respond to all guest complaints so he put away his still unfinished reports and left his desk to investigate further. He rode the elevator up to level nine and surveyed the empty corridor on that level. He was just about to head in the direction of the complainant's room when he spotted the door of room 967 nearby was ajar. He slowly pushed open the door and called out timidly.

'Is everything all right in there? I am the manager.' Getting no reply he switched on the light by the door and his eyes immediately fell on the pale figure lying on the bed, dead eyes staring blankly at the ceiling.

He sighed. It was going to be a long night.

The hotel manager picked up the phone on the now empty table and dialed the local police station. Once he had explained to the duty officer what the problem was, he placed a call to the medical examiner's office. It was hotel policy to always involve the medical examiner's office whenever there was an unexplained death at the hotel. This young assistant manager was not about to risk the reputation of his establishment, nor incur the wrath of his superiors by not following established procedures. He also knew his boss would want a full report the next morning.

Back at his desk, having touched nothing in room 967, except for the phone, he was once again leafing through his records when a police car and the medical examiner's station wagon both arrived simultaneously. Two policemen strolled into the lobby, along with the medical examiner, accompanied by his assistant.

The night manager gave them his passkey to room 967, which he had thoughtfully locked on his hurried departure from the room and wearily directed the posse of investigators to the elevators nearby.

Dr. George Gorrie was a dapper little man, dressed in a black double-breasted suit. He had a pair of bifocal glasses perched on the end of his inquisitive little nose. He was in fact Senior Medical Examiner for the city of Atlanta, a post he had held for many years, and in addition to his investigative duties there, he also taught forensic pathology at the University of Atlanta.

This week was a change from his routine because Gorrie was filling in as a substitute medical examiner for the Washington area. His good friend Dr. Payne, the resident Medical Examiner for this city, was away on a much-needed vacation.

On reaching room 967 the diminutive doctor started to check the position and state of the body, while both policemen began to search the room for possible clues. Dr. Gorrie was an experienced and thorough man and he surveyed the naked torso before him very carefully, before touching anything on and around the body.

He had immediately spotted the Medicalert chain that hung from the limp wrist of the still warm body on the bed and he quietly read the embossed red letters on the stainless steel bracelet.

Saying nothing to his assistant he took out his notebook and jotted down his finding that the dead man was a hemophiliac. Already Dr. Gorrie knew that this was not a simple case of coronary thrombosis.

His attention became focused on two light bruises at both sides of the dead man's jaw and as his intent gaze followed the pale flesh down the neckline he noted too the small puncture wound in the sallow skin.

'Another heart attack, eh Doc?' said the policemen. 'Sounds like he had a hooker in his room. Judging by what the night manager said. Guess he took on much more than he could handle at his age, eh Doc?'

The tired cop was trying to find a quick solution; he wanted to get back to the warmth of his secure little station as soon as possible. He had noted what the hotel manager had said about the recent disturbance -the naked girl running from the hotel...the strong smell of perfume in the room...His discovery of the discarded dress beside the bed...and the condom on the body, had clinched it as far as he was concerned.

Dr. Gorrie did not respond, nor did he get drawn into any discussion on the physical inadequacies of middle-aged men in general. Instead he gave curt instructions to his assistant, who was hovering nearby, to have the corpse transported to the downtown mortuary. There he could do a thorough investigation and post mortem.

Once back in their station wagon Dr. Gorrie quizzed his assistant. Gorrie sighed sadly as his young inexperienced helper reiterated the same theory that the policeman had so quickly seized upon.

'Well we shall see,' said the doctor enigmatically. 'I shall require a blood culture, a toxicology screen and a full coagulation profile.'

The assistant was puzzled but knew better than to argue. A few more days before his real boss, Dr. Payne, would be back. Then Dr. Gorrie could return to Atlanta and pester his own assistant with his seemingly unreasonable requests.

A toxicology screen and blood culture, were certainly routine tests to ask for, but a full coagulation profile was not. It would take him half a day to carry out this complex analysis, to determine if the dead man's clotting enzymes were normal or not.

The young assistant did not care if the corpse had been an excessive bleeder or one that had a tendency to clot abnormally...nor had he seen the Medicalert bracelet.

He continued to mumble under his breath as he dutifully but reluctantly wrote down the list of tests that Dr. Gorrie's had just requested.

His boss Dr Payne, would not have insisted on all this work...not for what seemed to him like a simple heart attack. Ah well, sighed the assistant quietly to himself, he would worry about that in the morning.

Right now he was tired and only interested in getting these tests logged in and then getting home to bed.

Chapter 9:

Dr. Christian Foster

Christian Foster was sweating heavily, not because he was sick but because he was in the middle of a heavy karate work out at his dojo in Atlanta, a place he went once or twice a week for his martial arts training.

Christian, or Chris, as he preferred to be called, had taken up karate several years earlier. It was a much needed diversion from his medical studies and even though he could only spare one or two evenings a week to this ever demanding activity he had been a dedicated student. The regular exercise had transformed him from a lank skinny teenager into a well proportioned young man with broad shoulders and muscles suited to his 6 foot frame.

His sensei at the dojo always demanded his best efforts and Chris had risen through the ranks to the point that he now sported a brown belt around his waist. There were others in the room, several of whom had joined the club at the same time as Chris, who now wore the coveted black belt, but his instructor Master Enoida, who held the rank of fifth degree black belt in Shotokan karate, was a wise old man. He fully understood that Chris had other important demands on his time so he had allowed him to train whenever he could, without pressuring him to continue on to his black belt.

Karate had certainly provided Chris with all the exercise that he had needed. It had kept him in excellent shape, physically and mentally.

As an added bonus Chris had found that the training at the dojo had also improved his mental concentration. Chris had been able to put in long hours at the university and maintain a high grade point average throughout his medical studies. Chris did not resent his black belted colleagues; most of them had regular daytime jobs and were thus able to spend more time training in order to maintain their higher status. They in turn were aware of Chris's demanding work schedule and even though they sparred with him intensely they did not take advantage of his limited training. Not that they could have, for Master

Enoida always watched over them like a hawk and maintained strict discipline.

Chris had enjoyed today's work out. He had initially sparred with two of the black belts who had both tested his defensive blocks to the limit. He had then performed an advanced kata in front of the whole class. The punches and kicks, delivered against several imaginary opponents who were all attacking him from different angles, had required all of his concentration and physical stamina.

Chris and his friends finally wrapped up their evening session with some calisthenics and though his blonde hair was matted with perspiration, his blue eyes sparkled and his body radiated a healthy glow.

Stripping off his white cotton uniform Chris stepped into a shower and as the powerful jet of hot water streamed across his face he reflected on his good fortune.

Chris had been raised in Duluth, a small rural suburb north of Atlanta, and although his father had died when Chris was a boy, he remembered him with fondness. His Dad had been the local family doctor who had taken care of all his sick patients in the area, but had been unable to cure his own diabetes. Eventually he had succumbed to the devastating complications of this chronic disease and when the illness had finally taken its toll on his kidneys and eyesight, he had been forced to retire from his practice. Following his Father's untimely death Chris's Mother had encouraged him to follow in his Fathers footsteps and she had made sure that money from her late husband's insurance had been earmarked for this purpose.

After graduating with honors from Duluth High School, Chris had been offered scholarships at several prestigious universities but he had thoughtfully chosen the nearby University of Atlanta. This way he could continue to live at home and thus minimize the financial burden on his Mother.

Despite the loss of his Father his childhood had been a happy one and he still remembered the pride on his Mother's face as she witnessed the graduation of her only child, near the top of his class at medical school. Her eyes had been moist with happiness when Chris had stepped up to the podium and was conferred with the degree of Doctor of Medicine. The only thing missing on that wonderful day had been the presence of his Father.

Although money had been tight throughout university Chris had always managed to find part time work during the semester breaks. He had made many good friends on campus. Despite his good looks and cheerful demeanor he had not had any serious relationships with the opposite sex during his five years of medical studies. There had been several pleasant, though brief liaisons, but his studies had always taken priority.

It was not until he had finally completed his degree that he had allowed himself time to pursue and finally capture the beautiful young student that he wanted. The one he had admired so often as she crossed the grassy acres between the auditorium and the halls of residence.

Chris smiled as he changed into a clean blue shirt and recalled how this lovely creature, which had been the object of desire for many of his fellow students, had led him on. It had been quite a chase before she had finally allowed him to take her out on their first date. It had been love at first sight but it was only after weeks of constant close company that had made them realize just how deeply they felt about each other.

Leaving the dojo, Chris waved off his colleagues who were trying to persuade him to join them for a drink at a nearby bar. Most of them knew that he would decline. They were jealously aware of where he was headed...the lucky sod was off to meet his lady love, and they were not surprised that he declined their less exciting offer.

Chris climbed into his battered old Pontiac convertible and smiled, his old banger was not the car of a successful young doctor, but then he wasn't successful yet. Plenty of time to acquire the trappings of wealth, once he had it, chuckled Chris to himself, as he sped off into the night with the wind whipping through his rapidly drying hair.

Within minutes he was in the downtown core of Atlanta where the tall dark skyscrapers loomed overhead like sentinels guarding the city, and he steered his car into the well-lit parking lot of the Peachtree Plaza hotel.

Checking his watch to make sure he was still on time, he saw that he still had a few minutes to spare so he stepped from the elevator on the ground floor and stopped by a long bank of pay phones on the wall. Chris called his home number to check for messages and as soon as he punched in the code that triggered the playback mode of his answering machine he immediately heard an old familiar voice.

'Hello Chris, this is Doc. Gorrie here. Please call me when you have a moment, I have a very interesting case that I would like to discuss with you.'

Chris was intrigued. It had been a while since he had heard from his old tutor and he grinned as he remembered the kindly professor who had so influenced his career. Chris had studied forensic pathology under Dr. Gorrie, who it turned out had been a good friend of his late Father. He had obviously been a close friend as Dr. Gorrie and acted more like a kindly uncle than a respected pathologist and professor of medicine.

Dr. Gorrie had persuaded Chris to continue his postgraduate studies at the university and go on to get his Ph.D. degree. He had also been influential in helping Chris obtain his current position at the Center for Disease Control in Atlanta, a position much sought after by all the new graduates, so Chris was very fortunate to have the backing of such a highly regarded mentor. Still intrigued by the cryptic message, Chris walked through the hotel lobby and entered an intimate cocktail lounge within the hotel. There he looked around until he spotted his girlfriend by the bar at the far end of the room.

Linda De Vaal was a vision of loveliness even from this distance. She was perched on a bar stool, her long shapely legs curled seductively around the stem of the stool, Dressed in a plain black velvet dress that did wonders for her figure, her shoulder length blonde hair was drawn back behind her ears.

Each ear sported a single diamond earring, the only jewellery that she wore. They were a tasteful symbol to the country of her birth. Linda had been born and raised in South Africa and the golden tan on her flawless skin was still testament to the many days she had spent on the beautiful beaches back home.

Following her graduation in medicine from the University of Cape Town, Linda had immigrated to America where she was now completing her postgraduate studies in Virology. Like Chris she also worked at one of the research laboratories at the Center for Disease Control in Atlanta.

As Chris gazed longingly at her across the room he noted that a young bartender was also smitten by her beauty and was obviously trying to chat her up. Chris quickly crossed the room. Other men feigned lack of interest when Chris reached her side and kissed her full on the lips. Linda smiled up at him lovingly and her red lips parted to reveal perfect white teeth, which were in exquisite contrast to her scarlet lips and tanned skin. Chris draped his arm around her shoulder possessively and drank in the familiar Estee Lauder perfume that she always wore and which always aroused him.

'You are obviously dressed for something more than Pizza' Chris said jokingly.

'Right on, Doctor,' said Linda, 'I fancy crayfish tonight and there is an expensive restaurant nearby.' 'Sounds good to me', said Chris, 'except they only serve lobster, not crayfish, in this town.'

Linda slapped him playfully on the shoulder as she untwined herself from the barstool and stood up so close to him that the heady perfume now aroused him still further. The hem of her short skirt had now slid back to a more modest position but the way that her dress now hugged her breasts and accentuated her shapely hips, only attracted renewed interest from all the men sitting in the bar.

Putting his arm protectively around her tiny waist, Chris led her proudly out of the bar, away from the stares of the lonely drinkers. They crossed the lobby of the hotel and entered a seafood restaurant that was an integral part of the hotel. They were met at the door by the tail-coated maitre d' who proudly selected two large menus from his desk and escorted them briskly to a table by the window.

Chris pretended to scan the menu carefully but he knew exactly what Linda was about to order. Linda wagged her finger at him playfully. Chris beckoned the wine waiter and ordered a bottle of Grunberger Stein, Linda's favorite wine. She was happy that it was available again.

For years it was banned, along with many other South African products, in protest at their apartheid policies. The wine was a pleasant reminder of the sunny days and times that she had been forced to leave behind.

Linda knew what effect a couple of glasses of this sweet heady wine would have on her, and so did Chris. As they sipped their wine together they both smiled. After they had tasted the wine they both ordered the grilled lobster and as they waited eagerly for their main course to arrive they munched through the crisp green salad that had been placed in front of them.

They ate slowly and contentedly as Chris recounted the intriguing phone message that he had received from Dr. Gorrie. Both of them shared the frustration of not being able to find out more until the following morning.

The lobster finally arrived at their table and they skewered the chunks of white seafood with their forks and saturated each piece in the large dish of butter that had been placed between them. Linda laughed as the rich yellow butter ran down Chris's chin.

The feast of lobster and the sweet wine soon turned their thoughts to the evening still ahead of them and all thoughts of Dr. Gorrie were temporarily forgotten.

Chapter 10:

Chris and Linda in Atlanta

They reached his apartment on the South side of the city and as they rode the elevator up to the second floor Linda embraced him so longingly so that Chris no longer had any doubts about what was to come. In the early days of their courtship Linda had resisted his intimate fumbling; she had not even allowed him to fondle her, but tonight was different and both of them knew it. Once inside the apartment Chris locked the door behind them and headed for the kitchen.

'Forget the coffee,' she said. Chris clicked on a light over the night table but Linda immediately turned it off again. Her shyness appealed to Chris and made her more desirable. In the darkened room he heard the sound of her zipper and then the rustle of her velvet dress as it slid to the floor.

She came to him, her hot moist lips passionately searching for his and as she came closer, the warmth of her body drew him into her. It was her firm breasts, pressing against his chest that prevented him from becoming completely absorbed within her. Chris caressed her back and hips tenderly as he traced first the silken bra and then the gossamer panties that she was still wearing. As she unbuttoned the front of his shirt he unclipped her bra and it fell lightly away.

For the very first time Chris experienced the excitement of her exposed breasts pressing into his own naked flesh. She slipped one of her hands down the front of his pants and her hand found his erectness.

Then in a sudden surprise move Linda opened up her mouth so that it now completely enveloped his lips in hers and the hot sensuous experience almost blew Chris' mind. In response Chris moved his hands down the back of her panties, grasped her soft curvy buttocks firmly and pulled her whole lower body into him.

Linda took her hand out from his pants and pressed her inner thighs so fiercely against his now erect penis that the material between them was hardly noticeable. Then, with a sudden sense of urgency, she pushed both his trousers and his underpants down to his ankles and forced them to the floor with a delicate flick of her foot. At the same time he slid her silk panties over her hips and as he bent down to extricate them from her ankles his lips brushed against the soft mass of curls between her legs. She moaned quietly at this intimate touch and then cried out wantonly as his questing mouth found her womanhood and his tongue darted between the innermost depths of her thighs.

She pulled him more urgently now toward the bed and as he sat down on the duvet she first nuzzled her head between his legs before taking his erect member full into her mouth. Chris cried out with the pleasure and the ecstasy of this new sensation as he rolled back on the bed and started to kiss and caress her entire naked body.

Finally when they could both hold back their passion no longer Chris turned over on to his back and pulled Linda on top of him so that her legs were now spread wide about his own.

Chris shuddered as her scorching hot breath seared his neck at exactly the same moment that his throbbing phallus slid smoothly into her warm innermost depths. They both continued to thrust slowly and gently together until an explosion of sensuousness made them both reach their climax in unison.

After a few minutes of holding each other very tightly they both lay back exhausted, the heavy perspiration glistening on their young firm bodies.

Linda was the first to speak. 'Guess you will have to marry me now,' she laughed. Chris laughed too, but he knew that in his heart this was indeed their destiny.

He then kissed her tenderly on the mouth and this spontaneous action spoke louder than any words. Linda now knew that this man in her arms would forever be at her side and with this comforting thought she fell into a deep and restful sleep.

Chapter 11:

Dr. Gorrie's Blood Sample

Chris Foster was the first one to wake up and it took him a few seconds to realize that he was not alone in his bed. As he looked across the pillow at Linda he was amazed at how wonderful she looked at this time in the morning. Her disheveled but still beautiful hair was splayed about her serene face and her lips were still red despite the lack of any lipstick.

Opening his eyes wider Chris also saw that the gyrations of the previous night had left the bed in such disarray that both her pink tipped breasts were exposed and were now peeking tantalizingly out at him from beneath the duvet. Propping himself up on his right elbow he gazed with awe at the exquisite shape of each of the milky white mounds of soft but firm flesh.

He had spent enough time in anatomy class to know that no woman was blessed with a perfectly matching pair of breasts but Linda's were as perfect as they could be. On top of each mound was a ripe pink nipple, like a cherry on a sinful ice-cream sundae. Overcome with desire he leaned forward and kissed the one closest to him. Linda awoke with a start.

She too was not used to sharing her bed with anyone and she shyly pulled the duvet close up to her neck before she realized exactly where she was and what was happening. Waking further she sleepily recognized Chris and as her lips opened into a broad smile he gave her another morning kiss.

'Time to get up, sleepy head', I'll make the coffee…the one we somehow forgot last night', he teased.

Linda blushed visibly as she remembered the events of the previous evening and wagged a stern finger at him mockingly.

'Yes, you put the coffee on Chris and then I can be first in the shower', she giggled modestly.

'How about we shower together and save hot water', Chris taunted, but Linda had already leapt from the bed, a sheet firmly gathered around her nakedness.

The shower door had clicked solidly behind her before Chris was able to make any further ribald suggestions.

Chris donned his tracksuit and sneakers as he headed into the kitchen and switched on the coffee maker. He opened up the refrigerator and checked it for supplies. Inside were a full carton of eggs and enough bacon for both of them but there was no bread so Chris slipped out of the apartment and ran down stairs to the street below.

His early morning jog took him to a neighborhood bakery that was already open and exuding the wonderful smell of freshly cooked bread. After chatting briefly with the amiable counter assistant Chris bought a dozen fresh croissants and then trotted back to his apartment.

Running back up the stairs he could hear Linda singing in the kitchen and he could smell bacon sizzling on the stove. Without disturbing her he dropped the bag of fresh croissants at her elbow and headed for the shower. By the time he emerged from the bathroom, shaved and changed into clean clothes, Linda had already laid out the freshly buttered croissants and was layering rich yellow scrambled eggs onto two plates. Standing nearby were two steaming mugs of freshly brewed coffee. Linking arms they picked up their plates and took them onto the balcony where they could enjoy their breakfast in the early morning sun.

As they munched their food hungrily they grinned at each other across the table, like two naughty children. The sudden shrill ring of a telephone in the kitchen then brought them both back to reality. Picking up the receiver Chris recognized Dr. Gorrie's voice.

'Hi Chris, hope I didn't wake you,' said the perky man.

'No, we have both been up for quite a while', said Chris, winking at Linda, who then wagged her finger mockingly back at him.

Before he could say any more Dr. Gorrie launched into an explanation of his current case as if he were lecturing a room full of neophyte medical students.

'Fifty five year old Caucasian male…found dead in his hotel room two nights ago…apparently death was due to a massive coronary thrombosis'.Chris noted Gorrie's emphasis on the word 'apparently' but did not interrupt him as he went on further.

'The deceased was a lab worker from Detroit...Visiting Washington for a few days, where I happened to be working. Semen found on his bedclothes indicates that he recently had intercourse...with person or persons unknown'.

Chris smiled at this last comment; it was typical of a man who worked so closely with the police.

'So what do you think Chris', demanded Dr. Gorrie.

'Just hold on a minute Dr. Gorrie, there must be more to this case or you would not be calling me.'.

Just then Chris reached for a pencil and paper...to jot down some notes...key notes he knew he would need.

'Oh yes', went on Gorrie, 'the laboratory tests we did yesterday showed that his blood contained malarial parasites...and he also tested positive for AIDS.

Dr. Gorrie had introduced this last statement almost as an afterthought rather than as a primary finding, this made Chris very suspicious but it also made him smile. He knew that his wily old friend was now having a bit of fun with him.

It was perfectly reasonable that his old professor would call him for a consultation about an HIV case. The professor knew Chris was doing a Ph. D. thesis on the new sub-types of virus associated with this disease.

This wasn't enough thought Chris, his old tutor was a brilliant but cunning man, he was also testing him. Gorrie had always made his students probe, probe, probe, until they had uncovered each and every diagnostic clue before allowing them to postulate a possible solution.

Chris was not going to be fooled by his old professor so he jotted down meaningful notes before responding. He also underlined some key points, in anticipation of the searching discussion that he knew would follow.

Linda had now joined him by the phone and was studying his untidy scrawl, noting the key points that he was underlining as he balanced the telephone receiver on his shoulder.

Laboratory worker...recent intercourse...malaria heterosexual...HIV positive.

When Dr. Gorrie concluded by saying that the dead man's extensive medical background indicated that he had never traveled to any tropical countries, Chris jotted on the pad…extensive medical background…and he underlined it several times before he posed his first question.

'How did you get an extensive medical history Sir, on a man that had only been in Washington a few days?'

'Oh we wired his doctor back in Detroit and they sent us his complete medical profile ', chuckled Dr. Gorrie.

Chris was more suspicious now, very few people had a complete medical history, not unless they were chronically sick and were under a program of active treatment. Knowing that the devious old professor was throwing out red herrings in order to fool him, Chris now took a more direct approach.

'Tell me first about the AIDS Sir, did your staff assess the level of viral particle in the blood sample'.

'Of course young man,' replied Dr. Gorrie, and the levels were extremely high…Consistent with a victim that has had the infection for several years'.

'Then his very extensive medical history confirmed a previously diagnosed HIV infection?' Chris asked.

'No, as a matter of fact at his last medical check up, just ten days ago…there were no signs of either HIV or Malaria', the mischievous old doctor chuckled.

Chris again underlined this point…medical check up ten days ago…no HIV or Malaria detected. Then he asked Gorrie a more pointed question.

'So do please tell me Professor, for what reason was this poor man attending a medical clinic?'

Dr. Gorrie giggled audibly at the other end of the phone, 'because he was a hemophiliac. He was there for his regular check up…And of course his Factor Eight injection '.

This last statement came across the phone line like a bolt from the blue. Chris circled the word hemophiliac on his scratch pad and drew an arrow connecting the word hemophiliac to the word thrombosis at the top of the page. Seeing this, Linda gave Chris a puzzled look.

There was no way that a man with hemophilia…a man who had an inborn tendency to bleed…Would die as a result of a clotting problem like thrombosis.

56

Before Chris could speak again, Linda jotted something on his pad. Chris bent closer to read her comment...She had written just one word...snakebite?

As he read the word aloud Dr. Gorrie also heard him.

'An interesting idea young man, it might indeed be possible for a man with this bleeding disorder to suffer a massive localized thrombosis...but very few snakes in the world possess venom with strong enough blood clotting enzymes to cause this.

However our recently deceased man had no signs of any cobra fangs on his skin...Although there was a single puncture wound at the base of his neck. He also had a cut on his finger, one that was several hours old that I don't think was related to his cause of death'.

Chris marveled at the doctor's extensive knowledge and then remembered that it was not for nothing that this man was such a respected forensic pathologist.

'It sounds like you have already figured out the cause of death Professor...was he administered some kind of powerful clotting agent...which of course would make it murder and not accidental death.'

'Excellent. Excellent young Chris, but I am still puzzled by the high levels of HIV, apparently none were present at his medical check up ten days ago.

The professor continued on before Chris could reply.

'Of course the laboratory test done at that time could have been wrong but I have sufficient respect for his particular laboratory in Detroit to actually doubt that, especially in a patient being monitored so regularly.

Even if he was infected with HIV, at the same time that he was murdered with the externally administered clotting agent, it still does not explain the very high level of virus in his blood at this time, does it Chris?'

Gorrie was now well within his area of expertise and so Chris now responded much more confidently.

'Of course there is always the possibility that his particular strain of immuno-deficiency virus is a new one, one that multiplies in the bloodstream much more rapidly, but I am still puzzled by the concomitant presence of malaria'.

'Yes, that's what I thought ', said Dr. Gorrie, 'that's why I would like to send you a sample of his blood, so that you can study it more closely in your much better equipped laboratory at the C.D.C.'

'Only too happy to oblige Sir, but don't send it over by courier, I will personally drop by this morning to pick it up, then I can start work on it right away'.

Chris was far too excited at the prospect of testing this unusual specimen to wait for it to come by courier.

'That's wonderful Chris,' said Gorrie', I may not be here myself but I'll leave the specimen for you in my laboratory and you can call me when you have more answers'.

Chris put down the phone and gave Linda a hug. 'You and I can work on this together darling, much more exciting than your boring work at Ostrich'.

Chris often teased her about the 'Epidemic Monitoring Unit' where she worked. To fellow workers her department was known by its acronym EMU but Chris always called it Ostrich. Chris had insisted that this was a more apt name because they buried their heads in the sand, with all the paperwork that they had to do.

He was of course joking; for he had tremendous respect for her credentials and the important work that she was engaged in. He also had a grudging admiration for the sophisticated equipment that she had access to.

That was another very good reason for enlisting her assistance on this extremely baffling case.

Chapter 12:

A Morning in Muskoka

Chris and Linda were finishing their breakfast in Atlanta at about the same time that Rick Harvey was planning his, at his cottage a thousand miles north in Muskoka. Rick had woken early to the sound of a loon calling across the lake and as he rolled out of bed, he saw the sun streaming over the verandah outside. It was going to be a hot, lazy day.

Feeling well rested but hungry he changed into shorts and drove into the nearby town of Bracebridge, where he picked up some supplies and a copy of the local newspaper. With most of the weekend still ahead of him he was relishing the idea of catching up on the Canadian news over breakfast. He sped back to the cottage without a care in the world.

After parking the car, Rick busied himself around the kitchen, cooking himself a massive plate of bacon, eggs and fried potatoes. Armed with this and a freshly brewed mug of coffee he sauntered out onto the sunny verandah and spread his newspaper out on the large pine table.

Rick liked to read the Canadian paper from cover to cover, especially the weekend edition, it always contained a lot more interesting articles than the newspapers back home. He seldom had time to read the Detroit daily papers during the week and they were so depressing, full of gruesome reports of murders, rapes and muggings, events that occurred all too often in the crime ridden city of Detroit.

Turning to the Lifestyle section of the *Toronto Star* his eyes picked up the headline. 'Bloodlines, a special report on the Krever Commission,' and he read through the article that followed with mounting concern.

The Krever Commission had been set up to investigate a tainted blood tragedy that had occurred when more than 2,000 people in Canada had been HIV infected from transfused blood. An estimated 12,000 others had also been infected with the potentially deadly Hepatitis C virus. Now, after a two-year federal inquiry, headed by Mr. Justice Horace Krever, the commission had heard medical experts warn that it was only a matter of time before Canada, and the world, would face a similar crisis, with some other blood borne disease.

The article went on to detail how the lives of people from every city and province in Canada, had been shattered by the tainted blood, they had so innocently received.

Particularly affected were the hemophiliacs in Canada who, because of their need for regular injections of life saving clotting concentrate, which was made by pooling the blood from as many as 20,000 donors, had been particularly at risk, because of their multiplied risk of exposure.

The clotting concentrate that had been the life-saver of the eighties was now killing recipients in the nineties. Rick dropped the newspaper with a start. He had always been concerned about the possibility of getting Hepatitis, especially as he was always working in and around hospitals. He often handled blood too, during the demonstration of his laboratory equipment.

For this reason he had applauded a new company policy of providing all employees with Hepatitis vaccine at no charge. Rick had immediately made an appointment to see his physician and then had the vaccine administered.

When he had claimed the cost on his weekly expense report, he had smugly reflected on how much safer he felt. Rick had never worn rubber gloves during his demonstrations; they somehow detracted from the focus of his presentations. It was not until this moment that he realized that the samples he had handled could have also been contaminated with the AIDS virus, as well as hepatitis.

How ironic, thought Rick, he took care to always wear a condom whenever he had sex, which was quite often as Rick met many willing women at hotels and bars on his travels. Rick laughed, none of his friends would ever believe that he had contracted AIDS from a laboratory sample; they would assume he had finally got his comeuppance from the life of debauchery that they thought he led.

As he slowly sipped his coffee Rick suddenly remembered his foolish action of the previous day. He had not only stolen a tube of blood from one of his clients, he had taken it with no thought of what hidden menace it might contain.

Leaping from the chair Rick sprang to the bedroom where he had flung his blazer the night before and rooted through all the pockets. He was now desperate to find the tube of blood that he had so foolishly hidden there. Failing to find it in any of his jacket pockets, his first reaction was one of fear. Fear that he might have dropped it on his way out of the hospital. If that were the case someone would surely have seen it…on the floor of the hospital…or in the car park.

My God thought Rick, if that had happened he would surely be exposed for his thoughtless crime. All kinds of concerns ran through his mind as he ran outside to search his car.

What if someone had seen him drop it…they would have already called his office…He would be in deep trouble when he got back to Detroit.

Rick's guilt ridden conscience was now working overtime…he pictured the confrontation with his boss…the shame of exposure…the severe reprimand, followed by instant dismissal from the company. Sweating profusely, Rick reached the car and was still thinking the worst as he searched frantically through the folds of the seats and on the floor of the car. In a final act of desperation, he bent down and looked under the front seat and saw the broken shards of glass.

The tube was crushed between the seat mechanism, broken and empty, but Rick still let out a long sigh of relief. At least he knew where the tube was and what had happened to it as he gingerly extracted the pieces of broken glass.

Rick smiled broadly now, it was after all only a minor inconvenience for him to go to his office on Monday and pick up a bottle of safe and uncontaminated control solution, which is what he should have done in the first place. This was nothing compared to what might have been and the fears he had felt moments earlier, were now forgotten.

He peered closely at the stain on the carpet under the seat; it was all that remained of his transgression. Knowing that it would not be noticed, he returned to the kitchen and dropped the remains of the now inconsequential little tube in the trash-can.

Somewhat relieved Rick returned to his newspaper and moved on to lighter reading and he was once again enjoying the ambience of his wonderful surroundings. Rick read on peacefully for a while until he heard the angry whine of a scavenging mosquito.

He hated these insects, they were the only blight on his otherwise perfect weekends at the cottage and he searched for the whereabouts of the airborne attacker. Rick had been bitten many times during previous stays at the cottage and although he knew they carried no disease, these bites had always raised big welts on his tender skin that were itchy and annoying.

As if to mirror his anger the whine of the mosquito rose to a high pitch as the filthy insect zoomed down onto his forearm. The whine abruptly stopped and the silence was deafening as its hungry nose sought a fleshy entry point. Its long spindly legs were caught in the coarse dark hairs on his arm and it teetered uncertainly on his skin.

Rick brought his other hand down with a resounding smack. The foul creature was squashed beyond recognition but it left a trail of blood on his arm, some its own…some from an earlier unwilling donor.

How lucky we are thought Rick that none of the mosquitoes in Canada carry malaria. There were many awful blood borne diseases overseas but in Canada he was much safer, the mosquitoes and black flies here were indeed a bloody nuisance, but that was about all. Rick was oblivious to his own unintended pun.

Tired of reading his newspaper in the warm sunshine Rick decided it was time to start preparing for a party.

It was Saturday and many of the other weekend cottage dwellers around him were starting to arrive. He would have no trouble attracting a few party-goers once they heard his loud music and knew there were free beer, free wine and free snacks to be had.

Throwing down the newspaper, Rick went inside and fetched his stereo and a few citronella candles from the kitchen and brought them out to the deck that surrounded his cottage.

The music from the stereo would soon attract fellow revelers to his party and the citronella candles would hopefully repel most of the hungry mosquitoes that were inevitably scavenging in the area.

Chapter 13:

The Virus Rears its Ugly Head

The traffic was still quite light as Chris and Linda drove to the main police headquarters in downtown Atlanta. Reaching the tall gray building with its intimidating tinted windows, they parked in the visitor's area and climbed the stairs to the reception area that guarded the entrance. In addition to watching over the doors to the police station this area also monitored access to the justice department and to the forensic laboratories, which were both conveniently located in this same building.

Chris was dressed casually in shorts, tee shirt and a pair of scuffed running shoes so the policeman on duty eyed him suspiciously until Chris showed his identification card. The towering beefy policeman rose to his feet smiling now and escorted them down a passageway to the forensic area where Dr. Gorrie's office and laboratories were located.

'The doctor is not in today but he left instructions with me personally regarding a parcel you are to collect', the man said, puffing himself up with importance.

As all three of them moved into the large forensic laboratory they noted that it was scrupulously clean, the fact that it was now deserted gave it an additional air of stark efficiency.

The policeman dutifully watched over Chris and Linda, and somehow kept both of them under constant surveillance. His eyes even followed Chris as he went and retrieved a parcel from the large refrigerator across the room.

It was unmistakably the right package, it was the only one sitting there on the top shelf and it had Chris's name printed on it, in clear bold type. As soon as Chris had the parcel, they all left the laboratory and returned to the reception desk, where the policeman insisted that that they sign his book to acknowledge receipt of the package.

Back in the car Chris handed it over to Linda who immediately wedged it safely between her knees. She knew that the sturdy package would contain enough absorbent material to guard against breakage. It would also contain a disinfectant capsule as an added protection against leakage of its potentially infectious contents during transit, but she still held it very tight.

The traffic was starting to build up as they drove the short distance from the downtown area of the city, to the place in the inner suburbs where both of them worked. Pulling up in the parking area of the impressive Atlanta Center for Disease Control, they both sensed the warm contrast of this building to the one they had just left.

This building was white, with clear bright windows, and the lawns between the buildings surrounding the main block were neatly trimmed around brightly colored flowerbeds. They both alighted from the car and ran up the steps that led into this warmly familiar building and again came upon a uniformed security guard. This one waved as he recognized them both instantly. The guard had been sitting at his desk reading a newspaper, and as Chris signed the registration book, he returned to his paper without asking Chris for his identification card.

Security was minimal in this building, all corridors leading from this central foyer were each blocked by heavily locked doors and only employees with appropriate pass-keys were able to gain access to their respective areas. Chris and Linda strolled casually along the air-conditioned corridor until they came to a door marked 'Dr. Chris Foster Virological Studies '.

At the door Chris inserted his plastic pass card into an oblong slot and as the heavy door slid open silently they both smelled the faint odor of disinfectant.

Passing through the outer laboratory they both headed directly to the room where Chris kept his precious research samples and the special equipment he used for his ongoing experiments.

Chris took the parcel that Linda had still been carrying for him and unwrapped it carefully on the desk. With the lid now removed Chris donned a pair of rubber gloves, but before he extracted the solitary tube from inside, he examined the contents closely to ensure that nothing was broken or had leaked inside the package.

Satisfied that all was in order he pulled out the solitary glass tube and set it down gently on the table. They could both see that it contained about ten milliliters of pale straw colored plasma. The wall of the tube was marked with an identification number, but there was no reference to the patients name on the sample itself.

This was standard procedure to protect the identity of the patient. All the details on the case, including the patient's name, were on a neatly typed card that Chris found in an envelope attached to the parcel. Even as Chris had been unwrapping the parcel Linda had gone to a drawer nearby and dug out a stainless steel rack. Then she had collected three clean test tubes and brought them to the area where Chris was working.

Chris had meanwhile put out an automatic Eppendorf pipette on the bench and she watched him arm it with a clean sterile tip before he aspirated plasma from Dr. Gorrie's specimen into each of the three smaller tubes.

Once this was done Chris stored the large tube, with its remaining plasma, back in his refrigerator nearby. He returned to the bench carrying a long plastic rack that contained twenty dropper bottles. Each of these bottles had bright, color-coded stoppers and Linda watched with admiration as he continued to set up his special equipment.

The plastic cassette containing colorless agar gel was something that she was very familiar with. It had been used for many years to separate human proteins into their various fractions by a process known as electrophoresis. Much more recently antigens and antibodies had been made to react together in the agar gel and the patterns they produced had provided researchers with clues regarding their specific identity.

All reactions in the gel were under the control of a small electrical current. When electricity passed through the cassette, the flow of electrons induced the antigens and antibodies to come together and react. Linda had used the technique many times herself but Chris had gone on to modify the technique further.

He had cleverly adapted the standard equipment so that it was now capable of identifying new sub types of HIV by reacting them with samples of HIV antibody. Antibody samples that Chris had personally collected and carefully blended to ensure their potency.

His equipment was now capable of identifying any new sub types of virus that Chris was investigating. The reactions of the viral particles he was testing against his own panel of antibodies, gave him a unique and definitive fingerprint. Chris had built up a unique collection of HIV antibodies and each of the types that he had identified was now represented in the twenty dropper bottles in his rack.

Linda watched as Chris deftly placed one drop from each coded vial of his carefully blended antibody into one side of the electrophoresis cassette. At the opposite end of the cassette he placed one drop of plasma from two of the tubes in the stainless steel rack.

These were identical samples of the specimen from Dr. Gorrie's victim, the one that was now the focus of their extreme and mounting interest. This done Chris switched on the current connected to the equipment and left it to migrate as he turned now to the last remaining sample in the stainless steel rack.

This one was destined for the electron microscope.

Linda had been dutifully silent throughout all this time and as Chris finished preparing the sample she smiled as she remembered how hard Chris had worked in the past. It had been no mean feat to justify the purchase of this expensive piece of equipment, which was essential to his research and so necessary for the satisfactory completion of his thesis.

With the embedded sample now in place in the receiver of the microscope, Chris turned on the power and the device started up with a barely audible hum.

The sample was drawn automatically into the optical chamber of the instrument, which then transmitted the amplified image to a computer. This enhanced the resolution still further before displaying the now greatly magnified contents of the sample onto a color monitor above them.

Chris focused the picture on the screen excitedly and after passing over several of the viral particles that were now clearly visible against the pale straw background, he zeroed in on one and brought it up to maximum clarity.

They both gazed in awe at the apparition on the screen, at the demon seed that stood revealed before them. The virus looked just like a coiled cobra, poised to strike. Linda had shrunk back visibly as she recalled an earlier moment in her childhood when she had last seen something like this. She had been visiting the snake park near

Pretoria and the head keeper there had insisted on showing off one of his deadliest captives to Linda and her parents.

Donning a pair of safety goggles he had climbed into the snake pit and teased one of the spitting cobras by poking it with a long stick. In response to this taunt the deadly snake had shot a stream of poisonous venom into the eyes of the keeper and only the goggles he wore had saved his sight. The accuracy and spite of the vile creature had left an indelible mark on Linda's young mind…she had suffered a fear of snakes and anything that resembled them, ever since.

Striving to put this fear behind her, Linda resorted to some weak humor in order to break the silence.

'Well it certainly is a 'heavy' Chris, isn't it?'

This was Linda's own pet name for the deadly HIV, Everyone else involved with the virus pronounced it 'Hivvy' but now Linda's South African pronunciation of the word seemed more appropriate to the menacing shape that now stared back at them from the screen.

'It may well indeed be a Heavy' said Chris, mocking her slant of speech, 'But right now I am much more interested in knowing just what type of heavy it is'.

'Well you will soon find out darling Chris, the results of your electrophoresis should be about ready by now'.

Chris did not respond immediately but continued to stare at the coiled cobra shape as he narrowed his eyes and tried to discern the precise configuration of the tightly coiled tails at its base.

'It's somehow different…yet it's also the same…the same as all the other strains I have seen', he hesitated.

It was Linda's turn to remain silent now for this was far and away beyond her level of expertise.

Tired of staring at the magnified abomination they both retreated to the outer laboratory where the electrophoresis was waiting for his interpretation.

As he swept his trained eye over the tell tale traces on the gel, his mouth gaped open in surprise as he made his next pronouncement.

'This type is unlike any HIV that I have ever seen before, in fact, it has not reacted with any of the special antibodies I have collected over the past two years'.

Linda showed her surprise, 'Does that mean that you've found an entirely new strain of virus Chris?'

'Only as far as my work goes Linda, I still need to check with other scientists engaged in similar lines of research, to verify that it is indeed a new strain…at least before I start writing it up for one of our prestigious medical journals', quipped Chris.'

How will you do that and how can you do it quickly Chris'. Linda was somewhat insistent now, as she did not want someone else stealing her boyfriend's glory.

She suddenly hit on a great idea and blurted out excitedly, 'Why don't you photograph it Chris, then we can send the micrograph to other research labs on my computer database'.

'What a terrific idea Linda, that way we can verify if it is indeed the first example of this particular sub type'. As he said this Chris silently thanked his lucky stars that he had thought to enlist the help of such a smart and capable ally.

Chris knew that Linda had been recruited by the Emergency Monitoring Unit because of her well recognized expertise in both virology and computer science. When she had first gone to work there, the computer data-base that she had inherited had only been capable of communicating with virology labs in North America and Western Europe.

This had severely limited her ability to track epidemics that occurred elsewhere in the world so her boss had been impressed when Linda had expanded the data-base to include additional virology tracking stations in Durban, Cape Town and Johannesburg. Thanks to her old contacts back in South Africa it had been easy for her to recruit them onto her database.

She had since gone on to extend the data base still further, now it included major centers in Zaire, Botswana and Tanzania. Linda's supervisor had been especially pleased at this accomplishment…this part of the world was where the new deadly epidemics, like Ebola and Marburg, were cropping up.

'Yes lets do it right now', said Chris, 'We can make a micrograph here and then take it over to your computer, but I'll also take the remainder of the sample…in case your equipment is incompatible and I have to prepare another sample using your even more fancy facilities' he joked.

The electron microscope was still on and already connected to the imaging equipment nearby so Chris ran off a picture of the virus. After retrieving all that remained of the plasma sample from Dr. Gorrie, Chris grabbed Linda's arm and hurried her off in the direction of her department.

Linda's facility was situated on the same campus, not too far away, so they got there on foot in a few minutes. Once inside her department Linda took over…This highly computerized and specialized area was of course her domain and she liked to be in control here.

As she inserted the micrograph into her own computer, they both breathed a sigh of relief as it was accepted.

The built in scanner immediately digitized the picture into a format suitable for transmission and sent it along to the modem. Linda's high-speed modem, capable of transmitting pictures to any corner of the world, now awaited her command. She pressed the button that sent it on its way and Chris marveled that this simple action would make the micrograph accessible to all centers on her database instantaneously.

'Well Linda there is nothing more we can do here now and I've just realized that in all this excitement we have forgotten to have lunch', added Chris hungrily.

'So we have, but not to worry Chris, I know delightful seafood restaurant nearby that serves great crab cakes'.

'Wonderful idea Linda, I will put what remains of the sample we got from Dr. Gorrie in your refrigerator, I am sure it will be perfectly safe there in the meantime'. As they left Linda's department, happily arm in arm, neither of them could yet know what importance this last act would have in the hours that followed.

It was early Sunday morning in Johannesburg and Commandant Joubert was feeling smug as he sat by his pool and awaited the arrival of his guests. Almost three days had gone by since the death of Sam Nthala at the Institute and the disposal of his body at a nearby crocodile farm would ensure that his disappearance continued to remain a complete mystery.

The coded message that he had just received from his BOSS agent in Washington had also brought a satisfying close to the events of the previous week. His missing blood sample been tracked from

Detroit to a hotel in Washington, where Devere had successfully recovered it. and then destroyed it by incineration.

Joubert had considered having the renegade tubes of blood shipped back to him via diplomatic pouch but had decided against it. Parcels sent under diplomatic immunity were indeed beyond the reach of any foreign inspection but he really had no need of the sample...he still had lots of the valuable material stored in both of his private laboratories.

The second part of Devere's message made Joubert smile. The hospital laboratory in Detroit...the location of his missing viral hybrid...and the medical technician who had tampered with it...had both been destroyed. Joubert was now satisfied that all sources of potential discovery and exposure in the United States had been closed. This pleasant thought was now interrupted by the arrival of some of his guests and he rose to greet the carefully selected group of people that were now heading across the lawn.

Those invited included a politically correct mix of both black and white people, not because he had softened his stance on equality but he did want to create an image of support for the current political dictate.

Joubert's guest list was drawn entirely from the professional sector as this was as far as he was prepared to go in demonstrating his facade of enlightenment. Black government workers and their wives, plus an equal proportion of white-collar workers and their wives. There was also a smattering of single professionals from both races, to apparently round out the mix.

Joubert enjoyed his first hour sitting by the pool, watching the young girls cavorting in the water. He had leered after the pretty white ones but he had looked with distaste at their black sisters. His distorted gaze focused only on the most obese of the black bathers and he sneered at their pendulous breasts and drooping abdomens.

Their pathetic efforts to emulate their more attractive white counterparts became more offensive to Joubert as his inbred racial prejudices started to resurface.

He had avoided drinking too much wine as he knew the effect it had when consumed under the hot African sun. Instead he had sipped slowly on a few bottles of Lion Lager. He also constantly supplemented his intake of alcohol with frequent snacks from the many dishes of food that were scattered around the pool.

Some of his guests were not so cautious, especially the younger women, and as their inhibitions dropped, so did the tops of their bathing costumes. By noon some of them had discarded their clothes altogether and were now writhing drunkenly together at the shallow end of the pool.

The stark nakedness of some of the ebony skinned women, especially the fat ones, filled Joubert with disgust. He had finally averted his eyes and made a mental note to have his black garden boy, who was also responsible for the pool, schedule a thorough cleaning of the pool water when the party was over.

Joubert finally tired of ogling at even the most attractive of the white girls, secretaries and nurses from the area who at least kept their figures firm by exercising regularly at the nearby Bryanston health club.

Finishing off his beer Joubert spoke to some of the more sober guests and suggested they abandon the wild party and accompany him to Sun City, for an evening of gambling and entertainment. About a dozen of them had agreed and they all left in a convoy of three cars, with Joubert himself, leading the way.

To Joubert's dismay two young girls insisted on going in his car and they babbled non-stop in his ear as he sped along the dusty road toward Bophuthatswana.

They also wanted his car radio on full blast and whenever he turned it down, the two giggling girls kept cranking it back up to full volume. The pounding rock music and their incessant laughter had almost become unbearable as they neared the end of their eighty-minute drive across the parched wasteland.

To Joubert's relief they finally reached Sun City, or 'Sin City', as many people called it...the magnificent hotel and gambling complex that had become the economic savior of the poor and barren homeland that they had just driven through.

The girls clapped with glee as they passed the entrance and immediately the bleak scrub of Bophuthatswana became a glitzy green oasis that rose decadently out of the desert.

It had taken vision and a great deal of money, much of it from undisclosed sources, to carve out this pleasure palace in the middle of nowhere.

Joubert's dusty convoy passed by three magnificent swimming pools and a championship golf course before they came upon an area dominated by huge carvings of African figures and wild animals. This was the fabled 'Lost City' that formed part of the centerpiece of the vast theme park and the high-rise luxury hotels that surrounded it.

As they came to a halt in front of the largest hotel a smart black attendant, dressed in an immaculately pressed safari suit, complete with tropical helmet, opened their car door and held out a hand for the keys.

This was more like it thought Joubert, they were only a few miles outside Johannesburg but already it was like stepping back in time…to the good old colonial days. The Sun City complex was completely multiracial but it was money that ruled here and only black professionals, who had plenty of cash to squander, were welcome at this still mainly white dominated playground. Leaving his car in safe hands Joubert strode through the hotel lobby and headed straight for the casino.

The giggling girls were still hanging on his arm but as soon as Joubert gave them a hand full of banknotes they cut loose and headed off to the slot machines.

Joubert bypassed all the roulette wheels and Blackjack tables along the way and seated himself at the high stakes poker table, at the far end of the neon lit room. Joubert liked to gamble, and he also liked to win, so his mood brightened visibly as he quickly accumulated a pile of high denomination poker chips. His judicious play and a strict avoidance of the flutes of champagne that were forever at his elbow were the keys to his success.

The champagne was always provided free of charge at the high stakes poker table but Joubert was more interested in the cards. He continually waved away the bevy of scantily clad, busty young waitresses who purposely kept up a non-stop supply of the bubbly beverage for any player who was foolish enough to forget about mental concentration.

The two giggly girls kept returning to his side but he kept them at bay with a constant supply of bank notes, which they in turn exchanged for coins to feed their ever-hungry one-armed bandits.

Finally, tired of their constant interruption, Joubert gave them a thousand Rand chip from his winning pile in front of him and whooping with delight they both wandered off into the lounge in search of younger company.

Joubert now settled back in his chair, he was up about twenty thousand Rand and was really enjoying himself. Suddenly a burly, sun burnt white man came and sat opposite him at the poker table and pulled out a large wad of Zimbabwean dollars which he cashed for high denomination poker chips.

Joubert took an instant dislike to this obviously wealthy but disgustingly flashy man. His short cropped hair and the heavy gold necklaces that hung ostentatiously around his sun-reddened neck were offensive to the military man. His hairless chest was also crassly exposed down to his navel and his tasteless floral shirt gaped open so that his flabby beer gut swelled out over his thick leather belt.

Joubert guessed he was a colonial tobacco farmer from Zimbabwe, judging by his accent and the currency that he had just exchanged, but his hands were soft and uncalloused. Joubert scowled at the man...this was no real farmer. He had obviously inherited vast tracts of land from his Rhodesian forebears, now he simply lived off the spoils of his inheritance and happily raped the land of his birth. As long as this poor excuse for a farmer continued to grease the palms of the corrupt black government that now ruled this once proud white colony, he would continue to live in idle luxury.

Joubert shivered at this last thought...This might one day be the fate of South Africa, and he would be forced to live alongside men like this. No way, thought Joubert, not if I can stop it...The fact that the man was as white as he was, completely escaped him.

The arrival of this man at the table completely changed both Joubert's mood and his luck. As the Rhodesian started to win heavily, Joubert started to lose heavily. Joubert compounded the problem by betting recklessly and soon his pile of chips, together with a large wad of notes from his pocket, were all sitting in front of his distasteful opponent.

Joubert did not like losing but he did know when to quit. He knew that he could not compete with this obviously wealthy fop so with a disgusted grunt Joubert slid from his chair and strode off toward the nearest outdoor pool.

Emerging into the bright sunshine he had to don his sunglasses before strolling along the lush grassy knoll that surrounded the dazzlingly bright blue pool.

Then he spotted a young black amazon, sunbathing topless…and on her own. Sauntering purposely in her direction he stopped only briefly, to pick up a cold lager from a refreshment cart. On reaching her side he brazenly sat down close beside her and as he peeled off his shirt he stared at her longingly.

She reminded him of Lazelia although her jet -black hair was shorter and her breasts were smaller. Joubert noted that her erect brown nipples were pointing at him provocatively and he felt his loins begin to stir.

Raising herself on one elbow the girl looked at him directly and without a trace of embarrassment at her obvious nakedness she smiled up at him and spoke.

'Hello', is this your first time at Sun City?'

'No, I have been here often, but the gambling tables are not with me today, so now I am out here looking for lady luck 'he responded with an obvious leer.

They were soon chatting together amiably and Joubert raked his eyes over her shapely young body as she told him that she was a dancer in the casino lounge. Joubert was not surprised, her shapely legs were proportionate but muscular and there was not an ounce of surplus fat on her body.

The girl could sense his lustful eyes, scanning her body from behind his darkened lenses, but she was not concerned. She knew she was in great physical shape and she also recognized Joubert as a man of status and power, two attributes that she was very much attracted.

After an hour of lying close together in the warm sun the girl suddenly sat up, re-attached her bikini top and suggested that Joubert buy her dinner in the lounge.

'How about dinner in my private suite instead',

Even though he did not yet have a room at the hotel he knew that he could soon get one.

The girl readily agreed and as they reached the hotel lobby she stopped off briefly in the ladies room to freshen up. As soon as she had left his side Joubert sidled up to the desk clerk and slid a credit card casually toward him. In minutes he had secured a large private suite on the tenth floor.

As the girl rejoined him at the elevator she was both unaware and uncaring that this man she had just met had only checked into the hotel moments earlier. Joubert continued to ogle the firm ebony body that was clearly and provocatively visible beneath her flimsy pink bikini as they rode the elevator to his room. Inside the room Joubert picked up the house phone, ordered dinner, and then continued their idle conversation until there was a knock at the door.

At Joubert's command to enter a black waiter came into the room pushing a folding table on wheels. As he began to open up the two leaves of the table and spread out the white linen tablecloth they could both hear the sizzling steaks that he was just about to produce from underneath the lower shelf of the trolley.

The waiter was skilled in his art and quickly uncovered two hot dishes from beneath two individual silver covers. With a professional flourish he set before them two huge rump steaks, each of them smothered in mushrooms, crayfish and a creamy monkey gland sauce. His job now completed, the waiter left the room, bowing slightly as he retreated backwards through the door. He had thought nothing of the mixed race couple alone together in the room he had just left.

The law banning intimate relationship between different races had only recently been rescinded in the rest of South Africa, but in Sun City, in the independent homeland of Bophuthatswana, that law had been ignored for many years. Joubert had still not asked his Amazonian companion what her name was nor had he bothered to ask what she wanted for dinner, but the girl had not complained. She recognized Joubert as a man of action and decisiveness and she was quite happy to eat the free meal he had ordered. As he opened up the magnum of champagne that had been placed in a silver bucket on the white tablecloth, they drank from the crystal flutes that Joubert had filled to the brim.

Munching through their bloody but perfectly cooked steaks they chatted loosely about the many attractions Sun City had to offer.

75

The girl had already noted that this unknown man had no luggage, nor were there any other signs of his occupancy in this three-room suite.

However she could care less, she knew that if she was nice to this man, and that included straight sex if he wanted it, she would be amply rewarded Her dancing in the lounge did not pay well and it was only one of the things she did to earn a living.

Once dinner was finished Joubert raised himself from the table and strode purposely across to the bathroom. The pretty young dancer smiled brazenly, as she knew that he was not going there to freshen up, she had already noted the absence of any personal toiletries in his bathroom, when she had peeked in the door earlier. She continued sipping her champagne until Joubert came back out of the bathroom but then she had to stifle a loud laugh when she saw how silly he looked.

Joubert had emerged from the bathroom with his powerful frame clad only in a pair of tiny black leather briefs. It was not until she saw the leather flail in his right hand that her smile faded and her bravado went out of the window.

'Oh No, I am not into any sadistic stuff, I think that I had better leave right now'.

Joubert was puzzled but he was not about to cool his ardor now, not for some silly black tart that was about to be well paid for her services. As he advanced toward her she backed away into the corner of the room and put her arms above her head, as if to protect her face from his anticipated strike.

What a silly woman, thought Joubert, as he moved closer toward her, does she not realize that I am the one that likes to be beaten.

The frightened woman suddenly found some courage and as he loomed over her, whip in hand, she darted from beneath him and ran out though the open windows onto the outdoor balcony.

It was already dark outside and he could hardly make out her silhouette as she cowered against the wrought iron balustrade that ringed the little balcony. As he moved even closer he could see the whites of her eyes as they rolled frighteningly in her head but the imminent danger of the moment was beyond the extent of his tolerance.

He had already lost a lot of money at poker, now this stupid woman was playing foolish games with him.

Finally his patience snapped and he strode toward her. Abject terror filled her eyes as she put one leg over the protective railings as if to warn him to come no further. Joubert had grown tired of the game and as he came closer he reached forward and callously gave a push.

Her teetering body spiraled in the air briefly as she toppled backwards over the railings. Her gaping mouth tried desperately to suck in air and it choked off the scream that was trying to form in her throat, as she fell quickly and silently to the ground below.

As she hit the concrete patio below with a sickening thud her broken body twitched only momentarily before her eyes stared unseeing at the blank sky above. As fresh red blood welled up inside her shattered frame it began to trickle from her nose and mouth.

Up above Joubert had heard her body hit the ground but he was not concerned, he placed no value on the life of this poor black trollop who had refused to do his bidding.

With four quick strides he was back in the bathroom and was hurriedly putting on his clothes. Joubert's only concern now was getting out of the room and leaving the hotel. As soon as he had finished dressing he caught the elevator down to the ground floor and checked out.

As he left the hotel he heard the buzz of activity on the patio. Someone had already discovered the broken body of the girl lying there, but Joubert just sauntered on by. He knew that the inefficient local police would take hours to arrive on the scene and they would have no way of knowing which of the many balconies she had fallen from.

One of the safari suited parking attendants saw Joubert heading for the car park and ran to get him his car. The poorly paid lackey was only interested in getting a tip as he brought the Commandants limousine up alongside him. He could see that the big white man was in a hurry so he slid quickly from the seat and paused only to grab the tip that was roughly slapped into his outstretched hand. Joubert took over the car and sped off into the darkness.

Commandant Joubert felt no remorse for the dead girl he had left behind him but he did want to get away fast and the needle of his speedometer was at its maximum until he reached the streetlights of the outer suburbs of Johannesburg.

Feeling more secure now that he was now back in his own domain he eased off the accelerator and turned his thoughts to all the money he had lost. He silently wished that it had been the arrogant white farmer that he had pushed from the balcony.

That white bastard had ruined his luck at the tables and that was what had ruined his day at Sun City.

Joubert visualized the sun baked farmer lying on the patio with his soft white hands covered in blood and knew that his emotions had to be satiated before he headed home. The party back at his villa would be over and his servant would have cleaned up, but Joubert did not want to be alone in the house with the smell of stale tobacco and beer. Joubert nose wrinkled at the thought, but the memory of alcohol suddenly made him feel thirsty and spotting a bar by the road he brought his car to a halt in a cloud of dust.

Once inside the little tavern he ignored all the other patrons as he sat alone at the dimly lit bar and ordered a bottle of brandy. Within an hour he had finished off the bottle and throughout this time no one had spoken to him. None of the other patrons were stupid enough to bother the big man at the bar who was obviously drinking away his problems and had no need of company. When the brandy bottle was finally empty, Joubert staggered from the bar and drove his car unsteadily along the deserted road to his house.

The house was in darkness as he staggered up the stairs and threw himself onto the bed and immediately fell into a shallow and very restless sleep. Throughout the night his brandy soaked brain dredged up old memories and took him back to the farm of his youth.

His Father had been quite poor but he had still been able to employ many black helpers. They were paid no wages but had to survive by growing maize on a piece of scrub-land that was grudgingly allocated to them.

It was a miserable existence but at that time it was all that the ruling white tribe of Afrikaner's would allow.

As a young child Joubert had learned from his Father that black workers were there to be bullied and abused. The fact that they were too afraid to complain or fight back only heightened the pleasure of mistreating them. He had also been taught, both at home and at church, that the Afrikaner tribe he was a privileged member of, had been chosen by God to rule all the black races of Africa.

This lesson had been so etched on Joubert's soul that any subsequent discussions on the dreadful inequity of apartheid would have been completely lost on him. To this day the only regret that he had ever had was the loss of privilege and unquestioning obedience that he had so enjoyed as a small child.

Just before he finally dozed off into a deeper sleep Joubert resolved to wake up the following morning with renewed zeal for his secret project.

Chapter 14:

Joubert's Virus Returns

It was Monday morning and Commandant Joubert had woken with a splitting headache and a mouth that felt like the bottom of a buzzard's cage.

His fierce hangover had forced him to forgo both his breakfast and his ritual morning swim before coming to work. He was now sitting at his personal computer terminal trying to pull himself together and focus on his work.

With a few taps on the keys he entered the secret password that allowed him access to the encrypted files that were stored on the hard drive and started browsing through some old files.

As soon as he came to a file coded 'Heartworm', he called up the results of an old experiment that he had carried out many months before. These were all secret experiments that he had carried out in an attempt to modify a virus that caused heartworm in dogs. A virus transmitted to dogs by the bite of a mosquito.

Joubert had successfully propagated the transmission of this disease into a number of laboratory animals, including some monkeys, but when he had tried to extend his study to include humans, the experiment had failed.

Despite Joubert's tenacity in increasing the frequency and strength of the injections, the poor black volunteer, who had no idea of what Joubert was doing, had died slowly and painfully of septic shock.

Joubert had covered his tracks well, one of his white guards had disposed of the body at the crocodile farm and there was no other record of this failed experiment. Realizing now that these files were of no further use he tapped on the delete key and erased them.

Reviewing this failed experiment had done nothing to raise his spirits but the memory of how he had callously disposed of the unfortunate human guinea pig gave him a feeling of omnipotence, a feeling he had not experienced since he had commanded his regiment on the Angolan border.

With the 'Heartworm' file safely erased he moved on to another file that was designated 'Malaria-HIV.' This time he smiled as he carefully avoided the delete key. This was his most successful experiment and because it was a testament to his brilliance he did not want to erase it. He knew that one day soon it would make him infamous.

Bored with browsing through his secret files Joubert decided to log on the main computer. He would use his modem to access the latest bulletins that had come into his virology department over the last few days. The screen in front of him flickered as his modem came upon a large graphic file waiting to be downloaded.

He had to wait a few minutes for the multi-megabyte file to be transcribed to his monitor. His screen started to clear as the incoming micrograph revealed itself and his mouth fell open as he recognized the familiar serpentine shape on his high-resolution monitor.

Joubert glowered and growled audibly, this was his baby and looking at it now was like looking at an intimate picture of his lover in someone else's wallet.

His mind searched frantically for answers, how could it be, how could this unique seed of his demented mind be now coming over his modem...from some unknown site far away?

His hands trembling with rage now, Joubert moved the mouse cursor onto the attachment icon and activated it. He had to find out who had transmitted this picture...an exact replica of the creation of his own cunning.

The screen now gave details of the case from which this virus had been recovered...a laboratory technician from Detroit...found dead in a hotel in Atlanta...Initial blood work up revealed Malarial parasites and H.I.V, and was this a possible new variant of H.I.V.?

A light suddenly went on in Joubert's brain. The report gave no details regarding the cause of death but Joubert was more interested in the names of the doctors who had worked on the case and who were now flashing this picture around the world, hoping to claim a new discovery. He found their names at the bottom of the page Gorrie...Foster...De Vaal. Joubert wrote all three names on the blotter on his desk.

As soon as he had calmed down he picked up the phone and dialed his agent in Washington. Without waiting for any explanation he blasted Devere at the other end for his incompetence in dealing with the missing tube of blood. He ended his message with some very explicit instructions.

'Eliminate all three...Gorrie...Foster...De Vaal, destroy all evidence in their possession...use any action that you deem is necessary'. Devere at the other end listened dutifully, he knew better than to interrupt his angry superior until he had finished.

'Yes Sir, I will deal with it immediately.'

As soon as Joubert replaced the receiver, the phone immediately rang with an incoming call and as he held the receiver to his ear he instantly recognized the guttural voice of his political supporter in Pretoria.

'What happened Joubert?' your Washington agent failed to carry out your orders', said the faceless one at the other end of the phone. Joubert was stunned, how could his master have known about this, how could he have found out about it so soon? Then he realized that either his home phone, or the one at the South African embassy in Washington was tapped.

Joubert was annoyed with himself for being so naive. He knew only too well that The Bureau of State Security kept a regular surveillance on everyone of interest to them. The long tentacles of BOSS were able to squirm under every rock and uncover every secret. Quietly resolving to be more careful regarding what he said over the phone, Joubert continued to listen dutifully to the man on the other end of the line. When the man had finally finished Joubert replied.

'I have already taken care of the matter...my man in Washington will deal with the situation immediately'.

'That's what you said the last time,' came the sarcastic response. 'You realize that we do not allow failures a second time, not even for officers at your level'. His tone was now much more ominous. 'This project is too important, I am holding you personally responsible Joubert...I will not allow our project to fail', he ended, as the phone went dead with a click.

Putting down his phone Joubert exploded with anger. 'Who does he think he is, how dare he refer to it as 'our project', how dare he infer that he played any part. This is my project...mine alone, and I will have all the glory of my own creation.'

Cooling down slightly, Joubert's senses were racing now but he resisted the urge to recall his agent in Washington. No, he thought, the man already had his orders and Joubert would not demean himself by repeating them, his agent knew exactly what to do and this time he had better not fail.

Instead of calling Washington, Joubert went to a freezer in another room and carefully extracted six tubes from within. Returning to his office he wrapped them in cotton wool and placed them carefully into his briefcase. He would take them home and store them in his own freezer.

Joubert had learned that it was always wise to have a contingency plan; especially one that nobody else knew about He closed the lid of his briefcase tightly on the innocent looking package within and returned to his desk. This time he was smiling once again.

Chapter 15:

Explosion in Atlanta

Captain Devere was still smarting from the verbal castigation he had just received over the phone and as he paced angrily around his plush office in Washington, he cursed the three people who had brought him disfavor with his Commandant. Now he had to return to Atlanta and deal with the three problems he had been given...Gorrie...De Vaal...Foster, but this time he would make no mistakes.

Devere's office at the South African Embassy in Washington had all the trappings appropriate for his cover as Assistant Cultural Attaché, but he had no interest in culture His office contained nothing of a cultural nature, just the sinister tools of his trade as a trained assassin and spy.

Outside the protective walls of the embassy his phony title gave him diplomatic immunity so that he could carry out his missions with relative freedom and complete anonymity. Devere unlocked his credenza and took out a square black briefcase, which was also locked.

Even though no one else in the embassy was allowed into his office he was still cautious and he kept all evidence of his true calling safely hidden under lock and key. Unlocking the briefcase he opened the lid and checked all the contents inside. In addition to a few clothes and toiletries, that he always kept packed ready for immediate departure, there was a small caliber revolver with a fitted silencer.

The bag also contained a hypodermic syringe and some vials of unopened thrombin, items leftover from his previous mission. Satisfied with the contents of his bag he crossed to the other side of the room where there was a safe on the wall. He spun the dials in several different directions until the tumblers fell into place and the thick metal door swung open. Reaching into its cavernous depths he pulled out three flat packs of plastic explosive and then reaching in again he brought out three mechanical timers.

Returning to his briefcase he packed them all within the folds of his spare clothes and then locked the case again. Equipped now with all the supplies that he might need for his new mission Devere picked up the briefcase and headed for his BMW in the underground car park of the embassy.

Twenty minutes later he had parked his car at Dulles airport and was at the ticket counter of United Airlines. The lady at the desk informed him that the next non-stop flight to Atlanta was not scheduled to depart for another two hours but he purchased a ticket anyway. As he checked his square briefcase with the counter assistant, she was too busy to ask him for his passport or any other identification, which was seldom required on domestic flights.

He also had no trouble passing through the security scanner on his way through to the departure lounge. Devere was only carrying a rolled newspaper now and his innocuous looking briefcase was safely on its way ahead of him. He was confident that its deadly contents would slip through the minor security checks that were only applied to a small percentage of the large volumes of checked baggage.

The trio of poorly paid security guards manning the scanner were too busy chatting about their weekend's exploits to spare more than a cursory glance at the apparently innocent traveler that was now passing between them.

Devere sought out and found the private lounge that his airline provided for its preferred customers, and to which his 'Business Class' ticket gave him privileged access. Once inside he ignored the bar, with its long rack of liquor and wine bottles, which the airline provided free of charge to its frequent flyers and business customers. Instead he poured himself a coffee and settled himself into one of the private phone booths.

He used the next hour wisely, first to phone directory assistance in Atlanta where he immediately got the business telephone numbers of Dr. Gorrie and Dr. Foster. He also got the home number of Dr. Linda De Vaal…But this time he also got her home address. This was indeed a stroke of luck, thought Devere, as he dialed Gorrie's business number, to probe further regarding his current whereabouts.

The female receptionist at the Forensic Laboratory in Atlanta informed him politely that the party he was seeking was currently away from his office. Routinely apologizing for Dr. Gorrie's absence she gave Devere the address of the institution and even went on to give him specific instructions on how to get there.

How kind, thought Devere, the American trait of helpfulness and courtesy, with absolutely no regard for security, made his task so much easier and he smiled as he pulled out a map of Atlanta from his inside pocket.

He found the locations on the map for each of the three quarries that he sought and marked them with a cross. He then coded each location with an identifying letter, G for Gorrie…D for De Vaal…F for Foster.

Devere had also wanted the home address of Dr. Foster but the receptionist at the Center for Disease Control had been so guarded that Devere had not pressed her too hard, he had not wanted to arouse her suspicion. Devere would find out Foster's home location once he got to Atlanta…maybe the girl De Vaal could help him with this, he thought sadistically.

Satisfied with his plan of action, Devere placed the map back in his top pocket and settled back to read his newspaper and await the departure of his plane. As he read he also surveyed the other waiting passengers in the lounge and as he watched them over the top of his newspaper he spotted a young man leering at him from across the room.

Devere resisted the temptation to smile back and he avoided direct eye contact. He had no time for this kind of attraction, his time was limited and his current mission was too important to allow him this kind of dalliance.

Devere's flight departure from Dulles airport was on time and uneventful and even the meal that was served to him in business class was pleasant. This was good mused Devere, he would not need to stop for dinner later and could begin to execute his murderous plan as soon as he reached Atlanta.

The rest of his trip passed quickly but by the time he had retrieved his locked briefcase from the carousel and cleared the terminal it was already turning dark outside. Devere hailed a taxi from the waiting rank of drivers and gave his selected driver the address of Dr. Gorrie's forensic laboratory in downtown Atlanta. Devere rode the cab in welcome silence as his driver concentrated on the busy evening traffic outside.

Opening his briefcase, Devere carefully connected one of the flat packs of plastic explosive to one of the mechanical timers and then stored the combination pack into a canvas pouch that was fastened around his waist. He also took out the gun, with its fitted silencer, and slid it snugly into his inside breast pocket. In fifty minutes they were in the downtown area of the city. Devere paid the cabby as he was dropped outside a building that was clearly signposted as the Central Police Station of Atlanta, and which fronted the forensic laboratories. After the cab left, Devere circled the entire perimeter of the building and noted every exit and access point.

The front of the building was well lit and Devere could make out the figure of the policeman on duty. He was clearly silhouetted against a bank of video screens that the man was monitoring and it gave out an eerie green glow from behind his darkly uniformed figure. Devere quietly traversed the back of the building for the second time and as he closed in on the only area that was not in darkness he peered in through the heavily barred window. Judging by the centrifuges and microscopes inside Devere could tell that it was a medical laboratory. Inside the room Devere could now make out the figure of an older man, with a scalpel in his hand, which was bending over a naked, partially dismembered torso.

Devere was now on full alert as he realized that this was one of the men that he come to find. This old man, dressed in a white coat, was obviously a pathologist for he was at this very moment conducting an autopsy. Devere smirked as he realized that this had to be Dr. Gorrie. Only a dedicated older pathologist would be working this late, younger assistants would be home with their families, not working in the laboratory at this time of night. Peering still deeper into the far reaches of the room Devere could now discern a neat row of filing cabinets at the far end of the neon lit room.

Realizing at once that these cabinets must contain records of all the patients that had been examined in this laboratory, Devere decided that these too must be destroyed.

Devere had been well trained in the art of demolition and the explosives he carried were all manufactured by experts. All his devices were designed to achieve maximum destructive power and to deliver their devastating force in the direction that he needed.

After carefully estimating the size of the room, he thoughtfully pulled out a second pack of plastic explosive from his briefcase. This he affixed to the primed pack that he had first taken from the canvas bag about his waist. He then attached the doubly deadly package to the base of the large window, with a piece of adhesive tape from his pocket.

Devere grunted with satisfaction as he picked up his briefcase and strolled nonchalantly away from the window, until he reached a point fifty yards away.

Devere pressed an electronic transmitter in his pocket.

Even he was surprised by the force of the explosion, which immediately took out all the windows in the laboratory. At exactly the same time a fireball engulfed the room inside and blew all the contents within, including the old pathologist and his precious records, to smithereens. Devere grinned as he walked away.

There would now be two dead men in the laboratory…the pathologist…and the body he had been working on. but there would not be enough pieces left of either of them to conduct any future post mortems.

As Devere moved further away from the conflagration that was now raging he did not pause to wait for the fire trucks to arrive, nor did he spare a thought for the policeman at the reception desk, who had also heard the explosion. The police officer who was now running toward the forensic laboratory, where the noise of the blast had come from and where the blank screen of one of his monitors now indicated that something was sadly wrong.

The video camera that had minutes before been recording the activity of Dr. Gorrie in his laboratory was now a molten mass of metal and was no longer capable of recording the raging fire that was at that moment consuming what little remained of the laboratory.

As Devere reached a main road, one that was some distance from the outer periphery of the burning building, he hailed a passing taxi. Devere gave the driver directions to drop him a short distance from the address of the next quarry he sought…Linda De Vaal.

It was a routine precaution; Devere did not want to give the taxi driver Linda's precise address. He did not want his movements traced later…besides the short walk would provide him with a bit of exercise, Devere thought smugly.

The cab driver dropped Devere at a location that was about two blocks from Linda De Vaal's apartment and as Devere strolled down the street he watched the shadows carefully. He had no reason to fear, this was not an area frequented by muggers and even if there had been one foolish enough to try to deprive him of his briefcase, Devere would have made short work of them with his pistol. As Devere strolled along he noted that he was in an area of large older homes and that many of them had been converted into multiple apartments.

At last he came to number 1250, which was the address that he had been given by the helpful telephone operator and he was pleased to see that each of the occupants within had indicated their precise location on a lighted console on the wall of the dark building.

'De Vaal L'. was listed as the occupant of apartment 2A but Devere ignored the bell push alongside Linda's name and used his credit card to slip the latch of the sturdy wooden door that led to the communal reception area inside.

Devere knew enough of American convention to know that apartment 2A would be on the second floor, so he bounded quietly up the carpeted staircase and again used his plastic card to pry the door marked 2A, at the top of the stairs. Devere slipped stealthily inside the room that was now in complete darkness.

He made a thorough search of the three rooms inside and then, when he was satisfied that he was the only occupant, he clicked on the light switch.

Another piece of luck thought Devere as he searched the apartment thoroughly and professionally.

There was nothing of any interest in the refrigerator, but he was not surprised, it would be unlikely that anyone would keep samples of blood in their apartment, but this time he was taking no chances. As he moved on into the bedroom he placed his briefcase on the floor and picked up a large photograph frame that was standing on the bedside table and stared at the picture. It was a large colored photograph of a fresh faced handsome young man and as Devere read the handwriting at the bottom of the photo he smiled as he read the telling inscription' To My Darling Linda, All My Love, Chris '.

Aha, thought Devere, this confirms that I am on target. It also alerted him to the fact that Linda and Chris were somehow intimately connected. Maybe I will get lucky again, he mused, and will be able to dispose of my last two problems in one fell swoop.

Moving to the telephone now, his eyes lit on a small mechanical directory close by, and as he flipped it open at the letter F he found a listing for Foster.

Just as he had expected, written against the name of Chris Foster, in a neat feminine hand, were two numbers. One was marked 'Karate Club', the other 'Home number', but neither had a listed address.

As Devere angrily pressed the rest of the letters on the mechanical index and browsed through the names and numbers listed under each letter, he noted several numbers that had South African area codes. Prefixes that he recognized immediately.

Ah well, thought Devere, as he now made the connection between the numbers in front of him and the unusual surname of his female quarry, this renegade South African lady would have to be persuaded to give him the information he sought, just as soon as she returned.

Almost as an afterthought Devere switched off the telephone answering machine standing nearby and as the little red 'ready' light went out he reached up to the room light above him and unscrewed the bulb from its socket. The room fell again into darkness, except for the slight glow coming from the partially closed door to the bedroom.

, Devere sank into an armchair in a dark corner of the room so that the chair was facing directly toward the closed outer door and Devere removed the gun from his pocket and placed it ready for action on his knee.

As he settled back in the darkness to await the return of Linda De Vaal he thought that maybe, just maybe, he might get Chris Foster too. Fifty-five minutes later Devere's body stiffened and he picked up the gun as he heard a key scratching in the lock. The door opened wide and he could clearly see a solitary female figure outlined in the doorway.

As soon as the woman entered the room and closed the door behind her, Devere heard the light switch being clicked up and down repeatedly, as the young woman tried unsuccessfully to turn on the light.

Sensing her frustration, and then her apparent acceptance of a burnt out light bulb, Devere allowed her to move into the center of the room before calling out to her ominously.

'Do not cry out, do not make a sound, there is a loaded gun pointed directly at your head.'

Linda froze in her tracks and did not make a sound.

'Go to the bedroom and do exactly as I tell you', Devere continued. The familiar accent of the man sitting in her chair, sitting there in the darkness of her home, was almost lost on Linda as terror welled up inside her.

Who was this familiar sounding but unknown man…and what was it that he wanted. Linda imagined the worst now. Had he broken in to her home to steal her prized possessions or had he broken in so that he could now rape and ravage her. With fear in her every step Linda went through to the bedroom and as she frighteningly turned to try and look at her unknown caller, the light in the bedroom gave her back a little of her composure.

'What do you want? If you have come here to steal something, take it now and please leave'.

The dark apparition rose from the chair, a gun now clearly in his hand, but he said nothing as he came through the bedroom doorway and moved closer toward her.

'My boyfriend will be here soon, from his Karate class', she added lamely, hoping that her use of the word Karate would somehow induce the man to leave.

Devere smiled. He was indeed hoping that Chris Foster would return, that would save him the trouble of seeking him out. If Chris were to return, Devere would be happy to deal with any puny kicks and punches that her pathetic boyfriend might throw at him. Devere had done martial arts training too but he had also learned that only a fool would go up against a man with a loaded gun, especially a crack shot like Devere.

As if to negate Linda's thinly veiled threat Devere pointed the gun just slightly to the left of Linda's head and squeezed the trigger so that the muffled weapon quietly coughed out its lethal lead projectile.

As the bullet whistled past Linda's ear and embedded itself into the wall behind her she choked off her latent scream…she had no doubts now regarding the resolve and the deadliness of this dreadful man that now stood before her.

Linda now meekly allowed herself to be trussed up in a chair in the bedroom and even when the electrical cable that Devere had ripped from the lamp, bit cruelly into her soft wrists she did not cry out with the pain. Her body trembled and her skin crawled as Devere's hands touched her body as he secured the restrictions around her waist.

Devere had no interest in her sexually, had it been Chris that he had been tying up; it might have been different, for Devere was only aroused by his own sex.

The only thing Devere wanted from Linda was information. 'Where are the blood samples and the records of the case that you transmitted over your computer to South Africa.'?

Linda's mind began to race, she did not want to put Chris in harms way, but she also knew better than to lie to this dangerous and determined man. She decided to give him some of the truth until her befuddled brain could figure out the reason why he was asking her these odd questions.

'If you mean the one that we transmitted the other day, they are now inside the Center for Disease Control, locked away safely where no one can get them', she added defiantly.

Building on her new found courage she went on.

'There is more of the sample safely locked away and guarded at the Forensic Laboratory, where even you cannot steal them'.

Devere laughed, 'I have already taken care of that sample, and the old doctor who started all this trouble'.

Linda gulped as her heart went out to the old professor. Dr. Gorrie was obviously in as much trouble as she was, maybe even worse, and the thought of this made tears well up into her lovely blue eyes.

Accepting that this frightened and weeping girl was too scared to be lying, Devere questioned her further.

'When is your boyfriend really coming home'?

Linda's instincts for the survival of Chris, the man she loved so dearly, cut in now.

'He is at his Karate class ', she insisted, 'but he will not come here tonight, he will go to his downtown laboratory when he has finished doing his workout'.

This last lie was Linda's attempt to put Chris safely among others, either at his Karate club or inside the place where he worked, safely out of this man's reach, she was also hoping that once this man left she could somehow free herself and phone Chris, to alert him to the danger that was now heading his way.

Devere found her story plausible but just to be sure he crossed to the phone and called Chris's home number. Hearing no reply from the other end, except for the click of an answering machine about to cut in, Devere returned to his briefcase and appeared to be getting ready to leave.

Sensing Linda's apparent relief at this move, he came back to her and took a long hypodermic syringe from his briefcase. He then slowly and deliberately filled it from the small vial that he now held in his hand. He waved the end of the needle so tantalizingly close to Linda's right eye that the ugly moistened tip brushed against her tear sodden eyelashes and she pressed her body into the back of the chair until her spine ached.

Noting her horror Devere put the syringe down on a table nearby, in a place where Linda could not avoid seeing it. He then pushed a dirty rag into her mouth and secured it with a strip of adhesive tape.

He brought his face so close to hers that she could smell his foul breath as he uttered his departing threat.

'If you are not telling the truth I will be back and my sharp little friend on the table over there will teach you a bloody lesson'

Devere laughed aloud at his own sick joke as he pointed at the syringe, then picking up his briefcase he was gone. Linda was left helplessly gagged and tied up in the chair but her eyes kept darting to the syringe that still lay there on the table.

It continued to taunt her as it sat there as a frightening reminder of the departing intruder's ominous threat.

Chapter 16:

Confrontation with Devere

Captain Devere walked briskly away from Linda De Vaal's apartment and on seeing a taxi, flagged it down. Devere gave the driver the address of his next destination…The Center for Disease Control.

The driver took off with his new passenger safely inside but showed no further interest in his passenger, nor did he wonder why he wanted to go to the government building at this time of night. He was just happy to get a fare. He didn't even check his mirror to see what his passenger was doing.

Under a blanket of darkness in the back of the taxi in the secure darkness of the cab, Devere delved into the depths of his briefcase and prepared another bomb, by attaching the last slab of plastic explosive to one of his two remaining timers. He then slipped the deadly combo into the canvas bag that was still secured around his waist.

On arrival at their destination, Devere got out of the cab at the front entrance of the Centre for Disease Control and after paying off the driver, waited on the sidewalk until the taxi had departed. As soon as Devere was satisfied that the cab was out of sight he started circling the building, checking the lie of the land, as he had done earlier at the downtown police station. He again noted a well lit reception area at the front, of the building except this time the man on duty was a civilian and there were no security monitors mounted on the console behind him.

As Devere skirted around the entire property he saw that all of the buildings in the complex were in complete darkness. He realized now that Linda had been lying about her boyfriend coming here tonight but it was no matter. Devere already had a better plan forming in his mind. After first hiding his briefcase under some thick bushes nearby, he returned to a public telephone kiosk that he had spotted earlier, at the front of the big building.

Inserting a quarter into the slot, he dialed Chris Foster.

This time the voice of a young man came onto the line, after the second ring, before the answering machine cut in. 'Chris Foster here, how can I help you', said the breezy young voice.

'Listen very carefully I will say this only once' Devere growled back into the receiver, and continued.

'I have your friend Linda hostage, do not do anything foolish and do not contact the police'. Devere went on, 'If you do not meet me in front of your place of work within the hour, you will never see your pretty girl friend alive again'.

The phone went dead before Chris could even splutter out a reply and although his mind was preoccupied with concern for Linda's safety, he realized that the rendezvous point that the mysterious caller had just given him, was at least a fifty minute drive away. Checking his wrist watch, Chris stopped only to slip on a pair of black leather clogs by the door and then, just as he was about to leave his apartment, he frantically ran back to the phone and dialed Linda's home number.

Her phone at the other end rang incessantly and when Linda's answering machine failed to cut in after twenty rings, Chris knew that something was wrong.

Slamming down the phone Chris raced out the door. Chris arrived at the Centre with only minutes to spare and as he parked his car he saw a tall dark stranger pacing up and down by the telephone kiosk in front of the building.

As Chris approached the stranger, the man turned to meet him and gestured ominously at a gun that was partially concealed by his jacket and which was now pointed directly at him.

'Do not try anything stupid, I want a guided tour of your laboratory...the life of your girlfriend depends entirely on your hospitality over the next few minutes', said the stranger, in a vaguely familiar accent.

'Where is Linda?' Chris spat, 'If you have harmed her in any way'...But his threat was left unfinished.

'Shut up and head for the door', the man was obviously not intimidated by Chris's veiled threat.

As they went in through the front entrance of the building, Chris was dismayed to see that the security guard was not at his desk.

Damn it thought Chris, the guard must have gone off on one of his coffee breaks.

The dangerous man behind him must have been watching the place and had timed their entrance right. With this small chance of distracting the gunman now gone Chris dutifully led the way to his laboratory.

As he reluctantly opened the door he tried to focus his thoughts on what his captor could really want, and more importantly, when he could overpower him and disarm him of his deadly weapon.

As the lights came on and they passed through the main laboratory Chris tried to gauge the precise distance between them and the exact position of the mans gun. Chris knew he would only have one chance to turn the tables but he passed on attempting something in the well-lit laboratory, instead he awaited further instructions.

'You have a sample of blood that I want, the one that you and your girlfriend transmitted over the computer…I want the sample and all the paperwork'…growled the stranger.

Chris was puzzled by this strange request but fear for Linda's safety and the sight of the gun, that was now out in the open and pointed at Chris's heart, left no time for doubting the strangers resolve. Crossing over to a large walk in refrigerator Chris pulled open one of the heavy doors and reaching inside, brought out a rack of coded tubes, each of them containing a small sample of plasma. Feigning fear and compliance, Chris took out one of the tubes and held it out dutifully.

As the man was about to grasp the tube, Chris let it slip through his fingers and the strangers gaze instinctively followed the downward trajectory of the little tube. Chris brought his foot around in a powerful roundhouse kick. As his foot found its mark his captors gun went spinning from his broken fingers and the man cursed as the tube and the gun fell to the floor.

Devere bent down swiftly to retrieve the gun but Chris moved even quicker toward him. The fingers on Chris's right hand was now taut and partially curved, so that it formed a hard fleshy striking blade, as he moved like lightning toward his distracted opponent.

The disarmed killer, anticipating Chris's knife hand strike to his throat, tucked his head into his chest to protect his windpipe. Chris saw the quick protective movement for what it was and immediately curled his fingers so that they strengthened his now rigid palm. As quick as a flash Chris brought the palm of his focused hand smartly upwards under the mans nose.

They both heard the crack as Devere's nose broke, but only Chris heard the second snap as Devere's long nasal bone was driven up into his brain. Devere fell backwards from the force of Chris's upward blow and his skull cracked on the door of the refrigerator, but this time he felt no pain. As the man lay prostrate on the ground Chris did not need any of his medical skill to know that the man was dead.

A wave of relief was quickly followed by a wave of remorse as Chris realized that both his oath to Hippocrates and his oath to his Karate sensei had both been broken. His oath to preserve life as a doctor and his promise never use his martial art skills in anger had both been overlooked. His concern for the safety of his beloved Linda, had temporarily transcended all this.

The sudden realization that he still did not know current whereabouts, or the welfare of his bride to be, made him so angry that he kicked the gun on the floor. The weapon spun away and as it glanced off the open refrigerator door it disappeared into the depths of the large freezer.

Still angry, Chris dragged the body of the dead man inside the refrigerator, so that his evil carcass now lay beside his equally offensive weapon. Chris bent down and quickly searched through the pockets of the still warm body, hoping to find some clue regarding Linda's whereabouts and some evidence of the stranger's true identity. In his haste to search him and then get on with the task of finding Linda, Chris missed the canvas bag that was still secured about the dead mans waist, beneath his shirt.

All he found was a brown leather wallet, which Chris slipped into his own pocket...he would examine it later. Finding nothing else of interest Chris turned away from the body, closed the large freezer door, and headed back to the reception desk.

The security guard was now back at his post but Chris had no time for pleasantries. He almost ran past him through the main door and just kept on going until he reached his waiting car. Neither Chris nor the security guard saw the rain-coated stranger lurking in the bushes but the stranger saw Chris. He wrote in his notebook...'Two men entered the building at 8.05 p.m. but only one-man came out again at 8.25 p.m.

Chris drove at breakneck speed back to Linda's apartment as this was the only place that he could think of looking for her and as soon as he reached her house he brought his car to a screeching halt. After unlocking the front door with the key that Linda had given him, he bounded up the stairs to her apartment. His urgent poundings on her door brought no reply so he stepped back, kicked the door forcefully until the doorframe splintered round the lock and the door flew open.

Chris fumbled impatiently with the light switch until he heard a muffled cry from the bedroom, and was overwhelmed with joy to find Linda inside. She was frightened but relatively unharmed. He kissed and hugged her anxiously as he ripped the filthy rag from her mouth and untied the restraints from around her wrists. Sobbing loudly, Linda blurted out the details of her ordeal.

Chris listened to her story, his eyes blazing with anger as she told him of the gunman's gloating over the apparent demise of his dear friend Dr. Gorrie. He only left her side briefly to telephone the downtown police station. On asking to speak to his friend, Dr. Gorrie, the policeman on duty sadly informed Chris that there had been a terrible explosion at the Forensic Laboratory, earlier that evening, and that poor Dr. Gorrie had been killed.

Tears welled up in Chris's eyes as he put down the phone and went to pour both of them a glass of brandy.

Bringing them over to Linda's side, he continued comforting her as he told her about the events that had gone on back at his laboratory. As he consoled her he suddenly realized how much he loved her and on how dangerously close he had come to losing her. Then in a moment of unabashed tenderness, he knelt down on one knee and looking up into her beautiful limpid eyes he asked her to marry him.

'Of course Darling Chris, of course I will marry you'. Then as an apparent humorous afterthought, which was Linda's way of dealing with tension, she added coyly 'But Chris, whatever will my parents say when they find out that I am marrying a fugitive from justice'.

'Gosh you are right', he said, responding to her jibe, 'I had completely forgotten that I have not yet called the police to report that awful mans death. In fact I called the police station just now and because of my concern for Dr. Gorrie, I forgot mention it…now I'm really in trouble'

'Hold on Chris', lets think this through carefully '. They fell silent for a few minutes until Linda spoke.

'You know that you have not done anything wrong Chris, there must be some way out of this'.

'What do you mean Linda', he said intently.

'You did say that you left his body in the fridge and that no one saw you going in together'.

'Yes', said Chris, immediately grasping onto her idea. 'If I go back to the laboratory now I can report his death as if it had only just happened'.

'That's right Chris, with his body still cold it will be difficult for anyone to pin point the exact time of his death, especially if you do it in the next hour'.

Neither of them voiced their sad simultaneous thought that it would not be Dr. Gorrie who would be doing this particular autopsy.

Stirred to action Chris gave Linda a long passionate kiss, left her apartment and then drove as fast as he dared, back again to the Centre for Disease Control.

As he drove he watched all his mirrors carefully, he did not want to get pulled over for speeding, for that would put an end to their clumsily concocted plan.

As Chris entered the reception area for the second time that evening he was greeted by a young security guard who had just come on duty and who was now starting his late shift. It was not surprising that he made no mention of Chris's earlier visit. He had not been on duty when Chris had last departed from this area but Chris still felt a tingle of fear as this time he signed the register.

It was apparent that no one had discovered the body yet, nor was anyone aware of the drama that had unfolded earlier in Chris's department. Chris walked down to his laboratory, as casually as he was able, and after gaining entrance with his passkey, he crossed to the large walk in refrigerator and swung open one of the double doors.

As he peered into the cavernous but well lit interior his mouth fell open.

The body was gone…and the gun was also missing.

Foolishly Chris searched within the refrigerator but there was absolutely no place for a body to be hidden. He then searched the outer laboratory, but here again there was no sign of the dead man or his nasty firearm.

As Chris scrabbled on the floor looking for clues, his searching fingers closed on the glass tube that he had earlier let slip purposely through his fingers and that had brought about his salvation. As Chris straightened up he automatically returned it to its protective place in the cold storage area. At least he knew that he had not imagined the whole episode but he was still frustrated that he still could not figure out what was going on.

With no evidence to show the police Chris decided against calling them. Instead he closed up the lab and decided to return to Linda, where he could tell her of this mysterious new development and where they could puzzle out this mystery together. As he passed by the security guard once more he could not resist asking the man if anything unusual had happened that night, even though he was dreading a positive answer.

'No everything has been quiet as usual Dr. Foster', came the reply as the young man swiveled back to the late night hockey game he had been watching on his portable television set.

When Chris got back to Linda he gave her a full account of his findings…or lack of them, in his refrigerator.

Linda threw up her arms in dismay and then shrugged her shoulders in resignation.

'This has been a long trying evening for both of us, why don't we both sleep on it and then my fiancée and I will work it all out in the morning', she said sensibly.

Chris smiled warmly at her reference to their recent engagement. At last he was absolutely sure that some of the events that evening had indeed happened and were not just a figment of his fertile imagination.

Chapter 17:

A Strange Dream

Chris Foster woke feeling as if he had hardly slept at all. His bed partner Linda had also been painfully aware of his restlessness, for even though she had held him tightly in her arms, he had tossed in his sleep and moaned the whole night through.

After they had showered and dressed Linda had made them both breakfast, now as they sipped their cups of freshly brewed coffee, they sat together in silence as they reflected on the strange events of the previous evening. Linda was the first to speak.

'Why don't we prepare a story board'.

'What do you mean', said Chris, perking up now.

'You know Chris, like they do in all the best detective shows on television. We can write up all the events and clues that we have so far and then add further ideas and theories as we develop them together'.

'What a great idea', said Chris, but Linda had already gone into the other room and emerged a few moments later carrying a flip chart and two black marking pens.

Propping the chart on the table in front of them Linda started to write 'Laboratory worker from Detroit' 'Died in Washington.' They both agreed that this should go at the top of the page as this event seemed to have been the beginning of all their problems.

For the next hour Chris and Linda took it in turns to jot down each clue and then they tried to arrange them in the order that they had occurred.

The death of the laboratory worker from Detroit had certainly seemed to have started the whole chain of events and was somehow connected to the appearance in Atlanta of the mystery gunman and the subsequent death of Dr. Gorrie. As they studied the flip chart Chris remembered the killer's wallet, which was still in his pocket, and he ran to the coat rack to retrieve it from his jacket pocket.

Bringing it to the table he placed it down triumphantly. At the same time Linda fetched the hypodermic syringe from the bedroom and placed it gingerly on the table alongside the wallet. These seemed to be the only two items that remained as a reminder of the gunman's presence in their lives. As Chris emptied out the wallet onto the table they were both disappointed to see that it contained very little in the way of personal effects.

Apart from a credit card that told them that the dead man's name was Carel Devere, there was a card that identified him as Assistant Cultural Secretary at the South African Embassy in Washington.

Linda's eyebrows arched visibly as she picked up the embossed card and turned it thoughtfully between her fingers, she knew now that the accent she had failed to recognize during her fearful encounter, was that of a fellow countryman. Then she wrote on the board 'Carel Devere…South African'.

Chris opened up a zippered compartment at the back of the wallet and pulled out several more business cards. Each of them bore the same name…'Carel Devere', but each of them indicated he had a number of different professions.

'It appears that our cultural assistant was also a computer salesman from San Diego, the President of an engineering company in Miami and also an accountant from Houston', said Chris disgustedly.

'Obviously all of these business cards are bogus and our deadly stranger adopted whatever role suited him for carrying out his dirty work', went on Chris.

'He was South African though,' said Linda. 'I was so terrified last night that I failed to recognize his accent. He has obviously lived in America a number of years, or he had cleverly learned to alter his Mother tongue. When he waved that awful syringe in my face, all I could think of was that I would never see you again'.

'Ah yes the syringe', said Chris, 'I have my own theory about what it contains and now I would like to confirm it, would you mind taking my blood sample'.

Linda was puzzled but she dutifully went and got her medical bag from the bedroom and placed it on the chair beside her. Opening up the black leather bag she took out a tourniquet and a plastic pack that contained a sterile syringe. Chris took the tourniquet from Linda, wound it tightly around his upper arm and clenched his fist so that the vein on his forearm bulged with the dark blood pulsating beneath his skin.

Linda took the clean syringe that she already had taken from its plastic wrapper and almost painlessly guided the sharp point of the needle into Chris's suffused vein. The clear plastic syringe sucked up a visible barrel full of Chris's venous blood.

'What are you proposing to do, 'said Linda as she inverted the syringe several times to mix the blood with the pre-measured amount of anticoagulant that the syringe had already contained.

'As you know Linda, the sample you have just drawn from my arm should not clot as the manufacture of that syringe had thoughtfully added some clot inhibiting chemical but now I am going to show you a little trick that I learned during my first year medical residency. A trick that makes anti-coagulated blood, clot again'.

Linda watched with fascination, Chris took the assassins syringe from the table and dribbled several drops of the clear liquid it contained into the sample of the blood that she had just with drawn from his arm. Within five seconds the red liquid that had once been Chris's free flowing body fluid became a solid and immovable blood clot.

'Just as I thought', said Chris, 'this syringe our assassin threatened you with contains Thrombin. Thank God he never injected you with it or you would be just as dead as the laboratory worker from Detroit' Linda's eyes opened wide now as she grasped the implication of Chris's experiment.

'So now we know exactly how he was killed', said Linda as she noted this new fact on the storyboard.

'But we still don't know why he was also infected with both HIV and Malaria'.

'Yes that is puzzling me too', replied Chris, 'I really need to get back to my own laboratory so that I can carry out more tests on that sample that Gorrie sent us'.

Linda could tell from Chris's dark frown that he was still not anxious to go back to his own laboratory, as despite his inherent curiosity, his enthusiasm was being dampened by his fear of re-discovering the assassins body in his refrigerator.

'I have a great idea', said Linda brightly, 'Why don't you take a few days off work and come and do your tests in my laboratory, I'm sure I could arrange it'.

Chris brightened visibly now, 'What a terrific idea Linda, the sample is already safely stored in your refrigerator and I'll then have the opportunity to use some of your fancy equipment'. 'Hold on though', went on Chris, 'I'll need to get some of my special antiserum first, as I will need it for my experiments'.

'No problem', said Linda, 'Just make out a list of what you need and I'll be happy to go and get it for you', said Linda, determined not to see his mood darken.

Quickly drawing up a list of his specific needs, Chris's spirits started to soar as he relished both the prospect of working in Linda's department and the deferment of the return to his own laboratory. He also relished the opportunity of applying all of his concentration to this nagging puzzle. Once Linda had left the apartment with his precious list Chris used the next half hour to make some phone calls.

His first call was to his secretary, where he left a message on her answering machine that he would be taking a few days vacation. His next call was to the forensic laboratory in Atlanta where he first of all commiserated with the still weepy secretary regarding the demise of poor Dr. Gorrie, but then drew a blank when he pressed her regarding the doctors records.

Not only had the laboratory been destroyed, but all the records that the meticulous doctor kept there, had also been burned in the aftermath of the explosion.

His next phone call to the forensic laboratory in Washington was more fruitful. Dr. Payne, the medical examiner there, consulted his personal patient records. He had already heard of Dr. Gorrie's death and after expressing his condolences, he was able to tell Chris the name of the dead laboratory worker from Detroit...Bill Williams, and the name of the place where he had worked-.the Detroit Municipal hospital.

Dr. Payne was apologetic, he did not have a copy of Dr. Gorrie's report on the case, as the doctor's untimely death had pre-empted both the completion of the report and the final conclusions of Dr. Gorrie's sadly unfinished investigation.

Thanking Dr. Payne for his help, Chris called the main number of the Detroit Municipal hospital and asked to speak to their hemophiliac clinic. On reaching that department, Chris identified himself as Dr. C. Foster from Atlanta and explained the reason for his inquiry.

'Yes we were very sorry to hear about the death of poor Mr. Williams, he was a popular worker here at the hospital and a regular patient at our clinic' said the receptionist sadly. When Chris went on to request that she send him a copy of Mr. Williams medical records she readily agreed and promised to courier them down to Atlanta, that same day. As Chris was about to thank her and put down the phone, the receptionist at the other end of the line still wanted to chat some more.

'It's very fortunate that Mr. Williams let us retain his records in our clinic and not up in his laboratory'. 'Why do you say that?', said Chris politely.

'Oh, haven't you heard, there was a dreadful explosion in his laboratory where he worked and both his office and all his patient records were destroyed'. 'Isn't it ironic?' she continued, 'If poor Mr. Williams had not died in that Washington hotel as he did, he would have been killed in his own laboratory'.

Chris put down the phone and walked slowly and thoughtfully to the storyboard. As he noted this new fact on the white sheet, he twisted the black marking pen in his fingers impatiently. He could hardly wait for Linda to return and tell her of this latest development. Another hour passed before Chris heard Linda's key in the door lock and as he ran to greet her he blurted out his news excitedly.

'Our mysterious gunman was not only responsible for the death of Dr. Gorrie and the destruction of his laboratory, but he also killed Mr. Williams...and he destroyed the laboratory where he worked in Detroit.'

'My God', said Linda, 'Then we are both lucky to be alive and you should stop feeling guilty that you were forced to kill that dreadful man'.

'Yes Linda but we still don't know where his body is'.

'I know', said Linda rolling her eyes, 'unless one of his cloak and dagger henchmen came and took it away. 'Linda shivered after this last remark as it dawned on her that if others were involved in this conspiracy then they were both still in great danger.

Chris sensed her new concern and spoke to her sternly 'Yes we must both continue to exercise extreme caution as we continue to investigate these murders together Linda', but as he spoke he instinctively put his arms around her to comfort her.

'I'm so glad you came up with the idea of us working together Linda, as I don't want you out of my sight. First thing tomorrow we must take all my stuff to your laboratory and then I can get started on solving this puzzle, that appears to be a conundrum in an enigma'.

'Well I'm glad to see that you haven't lost your sense of humor Chris,' and she smiled at his silly misquote.

Later that evening they both went to bed early, but even though they made passionate love together, Chris was still very tense and as he fell into a disturbed sleep he had what was to be a recurrent nightmare over the next three days.

As Chris slept fitfully he dreamed he was back in his laboratory, working at his bench when he heard a scratching noise coming from the large refrigerator across the room. In his dream he opened up the large steel door and as the automatic light came on he saw the dead gunman lying on the floor.

As Chris stared at the body, the dead man's eyes opened wide to reveal deep, dark empty sockets.

As he continued to gaze in horror the sockets became a mass of writhing motion as little worms mutated within and then spilled out onto the dead man's chest. As they slithered over the body these ugly apparitions started to grow in size so that they became snakes, snakes that assembled in a circle around the body, the snakes then began to move in a strange and mesmerizing dance.

As each snake continued to grow in size each of them took on the form of a cobra, but their lower bodies were multi-colored, which was in stark contrast to the darkness of their drab hooded heads.

Now moving in unison around the body their brightly colored bodies became segmented and separated from each individual viper's head. As each snake passed over the discarded tail of the creature ahead of it, they each absorbed their partner's body until they formed one large fused communal body.

The huge body was now one continuous writhing mass but it still had many heads, each of which swayed independently and spat out a stream of green venom.

Chris awoke with a start his body covered in sweat and he lay there in the darkness, trying to decipher the significance of his dream.

He knew that something deep in his subconscious mind was trying to tell him something. Some piece of knowledge rooted deep in his mind that was now trying to work its way up to the cognitive area of his brain.

The ugly beasts that had slithered from Devere's empty eye sockets had indeed looked like snakes but they had also reminded Chris of the ugly AIDS virus that had looked back at him so menacingly from the screen of his electron microscope.

He decided not to tell Linda of his dream, she was already paranoid about snakes and she had already been scared enough by events over the past few hours.

Chapter 18:

Joubert Prepares to Visit America

As Chris lay puzzling over his dream in Atlanta, Commandant Joubert was troubled too as he paced his office in Johannesburg, several thousand miles away. For the past two days Joubert had phoned the South African Embassy in Washington at least a dozen times, only to be given the same polite but empty response.

The receptionist at the embassy switchboard had each time politely but firmly informed him that Carel Devere was away on an assignment and had left no message regarding where he could be reached. Each time Joubert had called he had taken the precaution of phoning from a public phone box as he had no desire to alert anyone else to his growing concerns. He knew that both his phones at home and at work were bugged.

Forcing himself now to accept that something was radically wrong Joubert threw away all caution and dialed the number of the Washington embassy from the telephone on his desk. On being connected, he again heard the now familiar voice of the embassy receptionist as she again went into her routine reply.

'I'm sorry Sir, Mr. Devere is away on assignment and he did not leave a number where he could be reached'.

Joubert lost his patience and he bellowed down the phone. 'This is Commandant Joubert, his superior officer and I must get in touch with him immediately.'

The receptionist blanched...The anger of the man on the other end of the line was obvious but she felt somewhat protected by the distance between them.

'I am sorry Commandant but I am unable to help you...It is very unusual though...Mr. Devere usually calls in every day to pick up his messages...we are starting to worry about him too,' she went on feebly.

Joubert slammed down the receiver violently as he finally accepted the fact that Devere had screwed up. 'Bloody fool! Can't I rely on anyone these days?' he yelled aloud as he sadly reflected on how things had changed so radically for him in the newly liberated South Africa.

Gazing absent-mindedly out of the window Joubert cast his mind back to the good old days and started to daydream. In the past his orders would have been carried out without fail by his white subordinates. Even the black terrorists had learned to do his bidding…after a little persuasion. This last thought brought back to him one of his greatest pleasures.

Back in the good old days he had especially enjoyed administering 'Radio Moscow' to some of the black suspects arrested by his men and brought to him for special interrogation. This technique, which was a favorite with the South African police, involved the application of two electrodes, one to the poor victim's ear and other to his genitals. Then, when a powerful electrical current was applied to the electrodes, any information that the victim possessed was soon blurted out painfully and quickly. Joubert had even enjoyed applying this technique to one of his female suspects when a suffocating plastic bag around her neck had failed to make her talk.

Once the electrodes had been connected to her wrist and to her private parts; she had told him all that he wanted to know, between her screams of pain and the jolting of her frail body. Joubert's smile faded as he realized that all this was in the past now, unless he could do something to bring back his warped vision of what he considered to be happy times.

The urgent ringing of the telephone on his desk suddenly and abruptly disturbed his twisted thoughts.

Joubert already knew who the caller would be. The guttural voice of his mentor in Pretoria immediately confirmed his earlier suspicions that his phone was indeed under constant surveillance.

'Your agent in Washington has failed you Joubert, despite your misguided confidence in his abilities', the man at the other end of the phone sneered sarcastically.

'Yes, I am aware of the foul up and I am already making other arrangements to deal with the matter,'

'Joubert I am not interested in your plans, only your immediate action, so get yourself on the next available flight to America and then deal with this personally'. The emphasis he placed on the word 'personally' left no room for any further discussion. Joubert did not even have time to reply as the caller at the other end of the line slammed down the receiver with a bang.

Joubert was really pissed off, not only had he incurred the wrath of his anonymous benefactor but he could not call Devere in Washington to vent his spite on him. Joubert also did not relish the fact that he now had to make the long arduous trip to the United States.

Joubert had only been once before and he had hated the whole trip. The eighteen-hour flight from Johannesburg to New York, with only one brief stop for fuel, had left him tired, grumpy and haggard. Even his seat in the First Class cabin had been cramped and uncomfortable after the first few hours. Then just as he had started to doze, the plane landed at the Isla de Sol.

This god-forsaken island was the only re-fuelling station that South African Airways was able to use. All other airports in Africa had closed their landing rights to their national airline as protest to the then government's apartheid regime. Once down on the bleak landing strip all passengers had been forced to leave the plane during the time it had to be re-fuelled.

Joubert still remembered how the passengers had been herded, half-sleeping and jet lagged into a sleazy restaurant, just off the tarmac. He shuddered as he recalled the shabby cafe, where the inflated prices bore no relation to the quality of the food, or the level of service. Then on his arrival in New York, Joubert's sensibilities had been further incensed by the officious black customs officer who had took pleasure in harassing the apparently important businessman from South Africa. The black oaf had even had the audacity to search all his suitcases.

Joubert shuddered again as he recalled the dreadful trip. Garish neon signs everywhere…And worse still, the sight of black professionals, all well educated and well dressed, had been a constant reminder to Joubert of what South Africa might one day look like. This last final thought spurred Joubert into action as he remembered that the solution to this distasteful image now lay completely within his own grasp.

Calming down now he rationalized that it was indeed necessary that he make the trip to America. His mission was too important to the ongoing secrecy of his project, it could not be trusted to any lesser mortal.

Trusting no one, he phoned South African Airways himself. On inquiring of their flight schedules to the United States he was pleasantly surprised to learn that he now had the option of flying to New York via London, Paris or Madrid. At least the liberalized government had opened up more travel options, Joubert admitted to himself grudgingly.

A stop off in Europe was certainly preferable to the long haul over the Horn of Africa and he noted down the flight times and options on a blotter in front of him. He also noted the names of the three people he had written there earlier, 'Gorrie...De Vaal...Foster', and these three names he now committed to memory.

Sitting back in his plush leather armchair he carefully hatched a plan that would effectively deliver him into the midst of the three enemies he sought. He smiled as he noted that all three of his quarries resided in the Atlanta area, at least he would not have to chase them all over America. Clicking on his nearby computer, Joubert typed up a brief but very subtle letter to the Director of the Centre for Disease Control in Atlanta.

His letter stated that he would shortly be visiting important virology centers in the United States and as he would also be in the Atlanta area, he would appreciate a tour of their very prestigious facility. He went on to state that this was a vital academic trip and essential to his ongoing research into viral epidemics.

At the bottom of the letter he left enough space for his signature and printed his name and title 'Commandant Joubert, Director of Virology Research'. Loading his printer with a sheet of paper that was embossed impressively and headed 'The South African Institute of Medical Research', along with a flamboyant coat of arms, he transmitted the typed text to his printer. He sat back as the completed letter curled out of the laser printer that was standing at his elbow.

After proofreading the note and signing it with a flourish, he placed the letter in his Fax machine and transmitted it to the Centre of Disease Control in Atlanta. He smiled at the ease of which he had obtained their phone and Fax numbers from a virology reference manual in his bookcase. Within seconds the letter was across the

Atlantic and an efficient secretary called Miss Simpson, at the Center for Disease Control had taken it off her Fax machine and delivered it personally to her Director's office.

The Director, Dr. Graham Levine, a bespectacled, dapper man, had beamed as he read the important looking letter. 'Of Course, Of Course, we will be happy to extend our hospitality to a visiting fellow scientist, please fax him an immediate reply'.

Miss Simpson dutifully returned to her office and typed out a reply. In it she included specific directions on how to get to their facility and then faxed it back to the number that had been boldly printed on Joubert's incoming letter.

Within the hour Joubert had his reply and he grinned as he read the invitation.

His American counterparts were certainly less security conscious than he was, but then they did not have as much to hide as he did. Picking up the phone again, Joubert called Jan Smuts airport and booked himself a first class seat on the next available flight to New York, opting this time to go via London. He also made a reservation at the Dorchester hotel.

A stop off in England would break the monotony; he might even take time out for a little recreation in Soho. Joubert pulled out his black leather briefcase and stuffed into it a pack of explosives, a timer and a strange looking gun that he took from his desk drawer. The gun was made of a shiny ceramic material and although it did not have the accuracy or stopping power of his Glock .40 caliber pistol...It would serve his purpose.

He was not concerned about local airport security, his BOSS status would allow him to bypass the system in South Africa and the ceramic weapon would not be detected by even the very best security systems in London and New York.

The bomb and its timer had already been fitted with an electronic demagnetizer, one that was especially designed to fool the airport camera's and scanners that he would be forced to pass through on his long journey to Atlanta.

Picking up his lethal briefcase, Joubert headed out the laboratory and drove home. He still had to pack a few clothes and have a shower before his impending trip.

Chapter 19:

Chris identifies a new virus

It was Chris's second day of work at Linda's Emergency Monitoring Unit and he was really enjoying it. Linda's supervisor had extended him every courtesy and he had been allocated a small but well equipped laboratory for his use.

In addition to the very sophisticated equipment that was now at his disposal he also had access to an extensive computer library, one that could call up the very latest information on any medical topic of his choosing.

On his first day in the new department Chris was like a kid in a candy store. He had first played with all the fancy equipment and browsed through the massive computer database, looking at all the latest data on Ebola and Marburg virus that the EMU workers had accumulated over many years. After several hours Chris had finally forced himself to get back to his project at hand, this was all very interesting but it had little to do with his current problem so he had reluctantly turned off the computer and focused all his attention on his current problem.

Like any good scientist he had started by repeating all his experiments on the sample of blood that he had first received from Dr. Gorrie. Despite the new equipment he was now using his results again pointed to the same conclusion. The sample taken from the dead laboratory worker contained a strain of HIV that was unlike any Chris had seen before. He had also verified the presence of the malarial parasites.

This was a finding that still continued to bother him. An extensive medical history on the laboratory worker from Detroit had shown no previous evidence of this fairly rare parasite and he had never traveled to any countries where malaria was endemic.

In order to confirm his earlier suspicions Chris had also carried out a thrombin assay on the specimen only this time he had used the correct methodology, but here again the results had borne out the same conclusions as the crude technique he had used in Linda's apartment. The abnormally high level of thrombin in the dead man's blood, left no doubt in Chris's mind…the man had been murdered by the external administration of a powerful clotting enzyme.

Once all these tests were completed Chris spent the rest of the day purifying and concentrating the virus particles in the sample. This was in readiness for the exacting task, of mapping out the genetic sequence of this strange new virus.

Chris sat back in his chair briefly and gazed at the tubes. He was keen to get started on the genetic analysis but the recurrent nightmare that he had experienced yet again last night, continued to nag him. The dream obviously held some clue, but he still couldn't quite grasp it. Chris pondered how he could help his mind to focus in on the subconscious information he was still being taunted with.

Finally he hit on an idea, he would review all the latest findings on Auto Immune Disease on the computer database, and then perhaps some snippet of information would trigger his own subconscious memory database.

Sliding the computer keyboard out from beneath the desk, Chris tapped out the code to access the master database and then requested a search of all files relating to AIDS. He only had to wait a few seconds before the information started to scroll down the monitor in front of him and for the next hour he poured over the latest data.

Acquired Immuno-deficiency Syndrome, also known as AIDS had become one of the greatest health concerns since the epidemic of Tuberculosis in the nineteen hundreds. AIDS was thought to be caused by an infection of the Human Immuno-deficiency Virus, which caused the eventual break down of the victim's own immune system. The virus somehow destroyed the specialized white blood cells known as helper lymphocytes, and without these cells the immune system was unable to function properly.

Chris learned from the information in front of him that all the patients with AIDS were HIV positive but not all patients that were infected with HIV exhibited the signs and symptoms of AIDS. Chris made a note of this fact on a notepad resting on his knee and then continued reading.

Since the first case report in 1981, the Centre for Disease Control had recorded well over three thousand cases in people of all ages in the United States. In North and South America there were estimates that more than two million people were infected with the

HIV virus and the World Health Organization estimated that 8 to 10 million adults and 1 million children worldwide were now infected.

It was also estimated that by the year 2000, more than 40 million persons would be infected with the deadly virus. More than 90% of these cases would likely reside in the developing countries of Africa, Asia and Latin America. None of this information was new to Chris but it brought home to him the severity of the problem. Chris sat back in his chair for a moment and reflected on the need for action. Action needed to ensure the survival of mankind.

The word 'survival' triggered something in Chris's mind and as he focused his thoughts on this word he unconsciously selected another different menu marked 'incubation/survival' on the screen. The program automatically looked for additional specific information that his more selective search asked for.

The mode of infection seemed to be a key-determining factor in the actual incubation time of the deadly virus. The range of incubation times from contamination with needle sticks or blood transfusion was stated to be several months to several years, whereas homosexual contact could have an incubation time of anywhere from one to ten years.

Chris sat back again, none of this explained the case he was working on, his brow furrowed as he pondered. The dead laboratory worker had been heavily infected with HIV virus at the time of his autopsy and yet his medical history clearly indicated that at the time of his last medical examination, only two weeks earlier, no HIV virus had been detected in his blood.

Chris toyed with the idea that the murderer, that he now had no doubts was Devere, had infected Bill Williams with HIV at the same time that he had killed him with the thrombin. This made no sense at all, thought Chris, why infect the laboratory worker with HIV as well, the thrombin was lethal and Devere was obviously too smart to risk exposing and infecting himself when it wasn't really necessary.

Chris also doubted that the HIV virus could survive in such a strong concentration of thrombin so he pushed this theory to the back of his mind and returned to the computer.

The information on the screen was now detailing the physical characteristics of the virus and Chris went on to read that HIV was a Retrovirus and that its detection depended largely on the presence of antibodies. However searching for its presence by looking for its corresponding antibody was almost like looking for a man by searching for its shadow.

Fortunately great strides had been made recently using the new technique of Polymerase Chain Reaction, which amplified the genetic sequence of the virus so that even low levels of the virus could be detected. This was all well and good thought Chris, but it still did not explain why the deceased laboratory worker had no detectable levels of virus at his medical exam, yet just two weeks later, his body was riddled with it.

Tired of this train of thought he decided to look for actual cases that investigators around the world had already seen and he clicked on a menu that was designated 'HIV Investigations'.

This new computer search immediately accessed a new and very different section of its clinical database. A series of titles scrolled down the screen, each of them indicating a research article, the name of the author, together with a reference to the medical journal in which it had originally been published.

Chris read each one, looking for something of interest until he came to one titled 'The genetic sequence of HIV by Joubert M. D.; South African Journal of Virology'.

This one looks interesting thought Chris as he highlighted the reference with his mouse pointer and then clicked on the box marked 'print' in the top left window of his screen. Within seconds the powerful computer located the stored journal and then delivered Chris a hard copy of the article. As Chris pulled out the copy from his laser printer, he saw that the title page even had a picture of the author...Joubert M. D.

As Chris perused the summary he smiled as he read that the work on which this article had been based had led to the award of a gold medal for original research. Obviously a very prestigious award for this particular scientist in his country of origin...South Africa.

Chris read the complete article carefully and he was further impressed, Dr. Joubert had done a brilliant job of both identifying and categorizing the complete genetic sequence of the HIV virus. Chris had not until that moment known that this had been accomplished.

Sadly the computer back at his own laboratory only accessed medical journals from Europe and North America and Chris made a mental note to thank Linda for giving him the opportunity to use her database.

Chris also looked forward to telling her of his latest find. Who knows he thought, she might even have met this famous virologist and countryman of hers. Chris was so taken by the Joubert article that he read it through a second time and he was so impressed that he decided to search for any other articles that this obviously brilliant scientist might have written.

This time Chris searched the database using the author's name rather than the medical topic and as soon as he typed in the name 'Joubert' the screen flickered as it moved to another section of its database.

On the screen now were a number of articles, all of them attributed to Joubert and all of them published in South African medical journals...But one of them caught his immediate attention. Chris highlighted the new title...'The Genetic Makeup and DNA Sequence of the Malarial Parasite'...Chris then instructed the computer to print him an instant hard copy. Retrieving the article from his printer, Chris now read it in awe.

He had no idea that South African scientists were of such high caliber, or that they had access to such sophisticated technology. He would tease Linda later. Chris was puzzled though as to why a brilliant virologist would turn his attention from working on the HIV virus to something in a completely different field...Malaria.

As a virologist, Chris would have left this kind of work to his colleagues in Microbiology but perhaps the South African scientists were not as fussy when it came to crossing the boundaries of different medical specialties.

Feeling some eye fatigue Chris decided to break for lunch and rather than disturb Linda at her work he ambled down to the cafeteria on his own. Then after grabbing a clean tray he selected a salad sandwich and bottle of orange juice, which he then took to an empty table at the back of the busy room. He had only been seated there a few minutes when a young scientist, with thick black curly hair and a full beard, sat down at the table beside him.

Looking at the identification badge on the mans coat, Chris noted that his new dining companion was called Dr. Abraham Levine.

'Hi, Any relation to our Director', joked Chris.

'No...unfortunately not, the same religion but a different family', replied the stranger, flashing a smile.

'So what work are you involved in Dr. Levine', quizzed Chris as he started to bite into his sandwich.

'I'm on loan from Tel Aviv University, doing post doctoral work on your governments genome project'.

'Hey, that's pretty exciting stuff', yelped Chris, as he was well aware of the money that his U.S. government had allocated to this important new project.

It was rumored that over a billion dollars had already been earmarked to the mapping of the human genome.

It was a complex task involving the identification of millions of genetic sequences but it would someday lead to the identification and eventual eradication of many genetic disorders.

'How fascinating', Chris went on, realizing that this man at his table was one of the brightest in his field in order for him to have been chosen for this key project,

'You are just the person I need to talk to', said Chris.

'Glad to be of any help that I can', said the friendly young Israeli, now starting to eat his own sandwich.

Chris explained his work and outlined the problem he was currently battling with. After listening to Chris's detailed explanation of the investigations he had carried out, Abraham sat back in his chair and grinned.

'Sounds like you have also got an interesting piece of work on your hands, and one that is not unlike my own work, except yours has some forensic complications'.

Chris nodded and waited for his colleague to digest all the information he had given him as well as his lunch. At last Abraham finished his sandwich and spoke.

'Rather than looking for similarities between the HIV virus you have now and the ones that you have investigated in the past...look for what is different.'

'How do you mean', queried Chris.

'Well, if you first eliminate all the parts of the virus's genetic code that you have seen before and then just concentrate on the dissimilar chunks that remain, you may find it somewhat easier'

'Great idea' said Chris grateful for the simple but useful direction that his new colleague had given him. It also seemed to be the direction that his recent nightmares had been subtly edging him toward.

'Try this approach Chris and if you need any help with the identification of any sections that interest you, don't hesitate to ask. After all I do have access to all the finest equipment that your wealthy government could possibly purchase ', he replied grinning widely.

'I am sure that you deserve the best Abraham'.

'Yes I certainly do, and both it and myself are at your disposal', he concluded immodestly and rose to leave.

Chris shook his hand and scurried back to his lab, anxious now to pursue this new line of approach.

Over the next two hours Chris used his precious collection of restrictive enzymes to cut pieces of the HIV virus into specific segments, each of them having a unique DNA sequence. He then tested them against his own material. It was no surprise to Chris that most of them reacted, which confirmed that these sections of the virus were exactly the same as all the other types of HIV virus that he had seen before.

However there was one exception...one tiny strand of DNA that did not react with any of his highly selective sera. Despite repeating the experiment several times Chris still kept coming back to the same conclusion.

This was a piece of genetic material that he had never encountered before. This small strand of code was not only different, but its presence as an integral part of the virus he was studying probably gave it some unique traits that that were also unknown to him.

Extremely excited now Chris picked up the telephone and called Abraham...and blurted out his findings.

Abraham too became excited and told Chris to bring a sample of the genetic anomaly to him...immediately.

Precious sample in hand, Chris took the stairs two at a time as he headed off in the direction that Abraham had given him over the phone and as he reached the door of his laboratory, Abraham was there to greet him. He extended his hand excitedly to get the sample.

'We will soon have the identity of this little beast,' chuckled Abraham as he fed the sample into a very impressive looking gene decoding machine in his lab. The machine made no noise as it aspirated the sample into its inner reaction chamber and the computer screen only flashed out some inconsequential signals before the printer at its side fed out its calculations and conclusions.

'How very interesting', said Abraham as he expertly scanned the complex printout he was holding.

'What is it, what is it'. Said Chris, unable to contain his curiosity a moment longer.

'Are you sure you didn't contaminate the sample Chris', said Abraham, his face more serious now.

'No I am absolutely positive that I didn't. I purified the virus particles several times before I extracted the genetic material, why do you ask?', went on Chris.

'Well this piece of genetic material is not related to HIV at all…it is in fact a piece of genetic code from the malarial parasite', announced Abraham.

'As you know Chris the HIV virus is totally unrelated to the malarial parasite, so this is a strange finding'.

As Chris digested this amazing new fact a bright light went on in his brain. 'Not if someone had genetically engineered it that way', Chris shouted out triumphantly.

'Why would someone ever do that, it's not only extremely dangerous to modify a killer virus like HIV but it's also unethical and immoral', Abraham gravely went on. 'No one could possibly know what dreadful new traits this genetic manipulation could impart to this already deadly virus, or what effect it would have on its virulence, why the hell would anyone do such an incredibly stupid thing', snapped Abraham.

'I think I already know the answer to that' said Chris, 'I have just been reading two very interesting articles, one on the genetic sequence of HIV, the other on the malarial parasite…Both written by the same author'.

'Hold on a minute Chris, its illegal to modify a deadly virus as you know, any new strain could totally decimate our unprotected population…are you sure of your facts?'.

Chris went on to explain that he had always had a nagging fear that one day the HIV virus might somehow mutate itself so that it could be transmitted more widely. He also told Abraham that he had always thanked God that the virus was not transmitted by the mosquito like malaria was. He had never thought that someone might actually do it deliberately.

Abraham's eyes opened wide in horror as he listened.

'That would be devastating Chris, the incidence of HIV is already alarmingly high in Africa and that is also where they have millions of mosquitoes'.

'Yes, and if it ever found its way into our North American strain of mosquito we would have the same AIDS problem that they are now having in Africa'.

'Who would do such a thing and for what crazy reason?', frowned Abraham.

'I don't know the answer to that question yet, Abraham my friend, but I certainly do intend to find out'.

Chris quickly gathered his things and raced out of the laboratory in search of Linda. When he found her Chris relayed his latest findings and she immediately grasped the dreadful significance of his theory.

'If this virus has been genetically manipulated so that it can now be carried by a mosquito, then it can transmit H.I.V. to any victims that it then bites'.

'I've also got a another shock for you Chris, the man you think may be responsible, is someone I know…and from what I hear about him, he is crazy enough to do such a thing'

'My God', said Chris, 'then please tell me everything you know about this mad man'.

Linda went on to recount that during her residency at Johannesburg General Hospital she had heard strange whispers about the notorious Commandant Joubert. True, he was a brilliant virologist, she had heard him lecture at many important conventions in South Africa.

But despite his fame he had been forced to leave the hospital under suspicious circumstances. It was rumored that laboratory animals had been brutally mutilated during the time that Joubert had been on duty. Despite his adamant denials, he was the only one that could have been responsible and although no charges were ever brought against him, he left the hospital very suddenly. Later he had managed to secure an even better position at the Public Health Laboratories in Pretoria.

'But how could that possibly be', interrupted Chris.

'Chris, it is not what you know, but who you know, and that is especially true back home in South Africa, in fact that is one of the reasons that I finally left there'. Linda also went on to tell him that this man Joubert was certainly capable of this heinous act, she herself had seen pictures of the poor mutilated animals and she had also heard other stories of his cruelty in the army.

'Joubert is a powerful and dangerous man, we must be careful what we do', said Linda rolling her eyes now.

'Yes I hear you Linda, but he must be stopped. We must find out where he does his experiments and also where he keeps all his dangerous material. If we don't stop him quickly he could cause more deaths than there was during the great plague of Europe.

'They both sat in silence now, mulling over the full impact of Chris's last statement and what it meant to the world. They knew that they had to concoct a plan to somehow send Joubert...and his virus, back to hell.

Chapter 20:

Joubert calls on the C.D.C.

The Delta Airlines flight from New York to Atlanta was on time but when it came to a stop in front of the terminal its punctuality did nothing to appease the mood of one of its passengers. Commandant Joubert sat among the other high paying passengers in the first class section of the aircraft and scowled angrily.

As Joubert unbuckled his seat belt and prepared to disembark from the plane he reflected on his displeasure over the past twenty-four hours. The first part of his trip from Johannesburg to London, on South African Airways, had been comfortable enough, and he had slept away most of the journey.

The second leg of his transatlantic flight from London to New York had seemed longer because he had been nauseated by what he had witnessed along the way. By choice he had not stopped over in either London or New York as his initial impression of both cities had been tainted by his inbred racial intolerance.

At Heathrow airport in London his sensibilities had been stunned by the vast number of black people that were all about him in the terminal. Not just black porters and baggage handlers but also affluent businessmen, whose dress and status were much like his own. Many of them were of East Indian or Caribbean descent but to Joubert they were all black, as that was what he had been raised to believe…if you were not white then you were black…no other distinction was necessary as far as the racially prejudiced Commandant was concerned.

The culture shock that he felt on arrival at Heathrow made him abandon his planned stopover in London…it also fuelled his determination to complete his mission.

The modern world about him seemed to be going to hell in a hand basket and the sooner he got to Atlanta and eliminated all threats to his master plan, the better. These megalomanic thoughts had forced a cruel smile from his thin lips as he moved through London's busiest terminal and scanned the newspaper hoarding that he passed along the way.

The English papers were screaming out their objections to Germany's latest demand for the slaughter of all British cattle infected with Mad Cow Disease. Many thousands of English cows now had to be killed because of their suspected infection with Bovine Spongiform Encephalopathy. Joubert smiled, his German cousins were on the right track. The destruction of millions of tainted animals was of course justified...the sacrifice of one species to ensure the dominance of another, was a concept that he totally agreed with and this new thought reinforced in his mind the virtue of his own twisted master plan.

After hurriedly departing from Heathrow, Joubert had arrived at Kennedy International Airport in New York, where his discomfort had been still further aggravated. The throng of travelers there had been even more distasteful to him than the ones he had encountered in London. The proportion of black professionals as compared to white businessmen in the terminal had been even greater than it had been in London.

As he made his way through them he had been stunned, the non-white travelers here appeared even wealthier than their British counterparts.

Most of them wore dark conservative suits and dazzling white shirts, so that their crisp starched collars stood out like white beacons around their mahogany throats.

Like most American businessmen their trousers had been too short and revealed both their highly polished shoes, and an inappropriate expanse of stocking. The only difference between these educated and emancipated black Americans and their white counterparts had been the flashiness of their jewelry.

All of them wore expensive looking gold watches and bracelets about their wrists and many of them sported gold earrings in one ear. Some of them even had gold hoop earrings in both ears and Joubert had shuddered when he passed a young man who also had a ring through his nose.

'You can take the black man out of the jungle but you can't take the jungle out of the black man', Joubert had muttered to himself sarcastically as he had quickened his step to the departure terminal that would take him on to Atlanta. The deterioration of mankind had been evident all about him in the vast melting pot of New York, the sooner he could conclude his task in Atlanta and got back to his superficial white enclave in Johannesburg, the better he would feel.

On the last leg of his journey to Atlanta Joubert had buried his nose in his notebook, both to discourage any conversation with his fellow travelers and to focus his mind on the mission ahead of him.

Like Devere, the killer who had gone to Atlanta ahead of him, Joubert had used his time in transit at the airport to map out the location of his prey.

The telephone in the first class lounge had yielded the location of Dr. Gorrie and Linda De Vaal, thanks once again to the efficiency of the U. S. telephone exchange.

However, unlike Devere, Joubert had not telephoned these numbers, he did not want some nosy receptionist getting suspicious about his inquiries. Far better to go there in person, thought Joubert, as he lovingly patted the briefcase containing the explosives and his gun.

Arriving in Atlanta at last, Joubert had grabbed his overnight bag from the arrivals carousel and headed out the airport terminal, where he immediately found a waiting taxi. The weather outside was humid and sticky and he was relieved to find that the cab was air conditioned as he quickly jumped into the back seat. Leaning forward he told the driver to take him to the Forensic Laboratory in Atlanta.

The cab reached its destination and as it circled the building Joubert noted the extensive damage to the rear of the building and knew exactly what had happened. Devere had at least completed a part of his mission. Joubert instructed the cabby to wait at the front of the building and then went boldly up to the reception desk where he asked to see Dr. Gorrie. The policeman on duty had no reason to question the integrity of this smart looking visitor at his desk. The man was dressed in an expensive looking pin striped suit, and he was obviously planning to meet a fellow professional.

The policeman shook his head sadly as he now told the well dressed stranger that Dr. Gorrie had unfortunately been killed in an explosion a few days earlier. Joubert did his best to hide the smirk of satisfaction that crept across his face as he strode back to the waiting taxi.

Joubert searched his mind for answers, his seemingly incompetent henchman had at least destroyed this laboratory and eliminated Dr. Gorrie, but what had happened then?

Ah well, one down and two to go, thought Joubert as he gave the waiting taxi driver some new instructions.

'The Center for Disease Control and step on it driver'.

On reaching this new destination, Joubert paid off the taxi driver and strode up the stone steps at the front of the building. Pausing at the reception area he extracted a business card from his top pocket, and presented it with a flourish, to the young lady behind the desk. 'Commandant Joubert…I'm here to see the Director of your laboratories…he is expecting me I believe'.

The tall and obviously important visitor impressed the girl behind the desk. After checking the diary in front of her, she smiled warmly before ringing through to the office of Dr. Levine, to announce Joubert's arrival. Joubert had been standing in the reception area for only a few moments when Dr. Levine came rushing out, his hand outstretched in a warm welcome.

'Do please come in Dr. Joubert, we are so happy to see you and have you tour our facilities,' gushed the diminutive scientist.

'Thank You Dr. Levine, I am grateful that you were able to accommodate my request at such short notice.'

'Only too happy to oblige a fellow scientist, I hope you had a pleasant journey…where would you like to begin,' the man went on without pausing for breath.

Joubert smiled as his mood brightened. He was now among fellow scientists and the disturbing images from the airport were now well beyond these hallowed halls.

'I am currently directing research into newly emerging strains of HIV, whatever similar research you are doing here at your prestigious facility, would be of interest,' Joubert's smile widened as he flatteringly ingratiated himself still further with Dr. Levine.

'In that case I will give you the grand tour of our entire facility first and then finish up in Dr. Fosters department. You will find that most interesting as he is also conducting research into new strains of HIV Unfortunately he is not at work today but I will be happy to try and explain what he is doing and show you around his laboratory', beamed back Dr. Levine.

Joubert stiffened at the mention of Dr. Foster's name but he maintained his composure as they both set off on their tour. As they passed from room to room Joubert feigned interest in each area as he cunningly probed Dr. Levine for the information that he was really seeking.

'I'm sorry I will not get to meet Dr. Foster, his work sounds very interesting…is he away on vacation?

'No,' responded Dr. Levine, 'I think he is doing some collaborative work with his fiancée, but I'm not sure where, probably in another section of our campus'. Levine carried on chatting as they continued their tour. 'Actually Foster's fiancée is also from South Africa, her name is Dr. Linda De Vaal, do you perhaps know her', went on the innocent Director.

'No I don't think so,' lied Joubert, his excitement rising on hearing the name of the quarry he sought.

He was still puzzled why Devere had not killed both De Vaal and Foster and he was annoyed that he still did not know what had happened to his missing agent. Two hours went by before their guided tour of the main building came to an end and Joubert was still doing his very best to look interested and attentive. The Centre for Disease Control was impressive but Joubert was far too obsessed with his real mission to take in all the sights, but he had asked pertinent questions along the way to maintain his cover.

When at last they arrived at the department where Dr. Christian Foster usually worked Joubert's dark eyes flickered brightly as he took in every detail. The work being done here and the actual lay out of this particular laboratory were indeed of great interest to him.

Joubert was especially interested in the serum samples that were stored in the large walk in refrigerator but when Dr. Levine swung open the door, Joubert was pretending to study the temperature control mechanism and the thickness of the insulated door.

'This is exactly the equipment we need back in my laboratory, do you mind if I take down some pertinent notes regarding the specifications', smiled Joubert.

'Not at all Dr. Joubert, would you like me to find you a vacant office where you can work undisturbed,' said the ever-accommodating Dr. Levine.'

No, I can do it right here…if that's O. K. with you Dr. Levine, I have my notebook and briefcase with me.'

Without waiting for a reply Joubert sat himself down on a chair close to the walk in refrigerator and pretended to search his briefcase for his notebook.

'That will be perfectly all right,' beamed Levine…'Actually I can go back to my office and catch up on some of my urgent paperwork, that is if you don't mind me leaving you here alone'.

Joubert could hardly believe his luck at being left alone in this particular department but his smile faded when Dr. Levine pressed a buzzer located on the wall and a uniformed security guard came striding into the room. Joubert cursed under his breath, this was obviously standard procedure whenever a visitor was left alone in one of the more sensitive areas of the Institute.

Joubert need not have worried; the young security guard who was supposed to be watching over him was far more interested in looking out of the window.

It was lunchtime and many of the pretty young girls who worked at the Centre were now streaming out of their various departments, heading for picnic tables outside. As they sat bronzing their nubile bodies in the warm sunshine, their shapely forms were of far more interest to the young security guard now watching them. By now he had edged his way so close to the window that his nose was almost pressed up against the glass.

Joubert took advantage of this timely distraction. While still pretending to take some notepaper from his briefcase, he quickly took the time bomb from inside his briefcase and placed it on his knee under the table. The security guard was still too busy ogling the girls outside as Joubert set the timer and fixed the deadly device under the table he was sitting at.

The guard had seen none of this and the ledge around the table now concealed both the explosive device and the adhesive tape that now held it firmly in place. Snapping his briefcase shut with a loud click, Joubert rose from his chair and announced aloud…

'All finished here, we can now rejoin your director.'

The young guard tore himself reluctantly away from the window and followed Joubert. He had a boring job and apart from the pretty young girls who worked on campus there was nothing of interest or excitement for him here. He sighed as he resigned himself to escorting Joubert back to the main office and getting back to his television in the restroom. Even before the guard got back to his own area Joubert was already thanking Dr. Levine for his hospitality and taking his leave from the accommodating old Director.

'When do you leave Atlanta, do we have time for dinner together tonight,' said Levine, eager to impress his foreign visitor with some American hospitality.

'Unfortunately not Dr. Levine, I'm catching an evening flight from Atlanta so that I can connect with an overnight flight from New York and on to London.' It was unlike him to give out this kind of information but he was smug after planting the bomb so easily. Joubert was also anxious to get out of the building and get far away from this fawning little man.

'Ah well, some other time,' gushed Levine, but he was also secretly relieved that he did not have to spend the evening with this foreign visitor. He had found Joubert a difficult man to talk to and he did have a lot of work to catch up on, before heading home to his family.

Fortunately Levine was unaware of what lay ahead of him, that evening…he had no way of knowing that he would not be seeing his wife and family ever again.

Joubert left Levine's office and as he passed by the receptionist he stopped to thank her for her help and hospitality. The warmth of his farewell surprised the young girl and she blushed. Not only was she caught unawares by his smarmy approach, she almost got caught eavesdropping. She too was bored with her humdrum job at the switchboard so she often left the intercom switch on, so that she could listen in on the director's conversation with his visitors. Joubert had approached her desk so stealthily that she had almost got caught red handed.

The receptionist was an avid traveler and she had been intrigued by the arrival of the tall man from Africa, a continent she had yet to visit. She had eavesdropped on the details of his return journey, New York…London, Johannesburg, with a mixture of admiration and envy.

This man in front of her was now heading to London, a place that she had always wanted to visit, but she wisely made no mention of this snippet of information, that she had overheard, as he posed his next question.

'Dr. Levine happened to mention to me that a South African friend of mine...Linda De Vaal, is engaged to one of your doctors...do you happen to know if they are out of town'.

The naive receptionist was happy to be of help to this important visiting dignitary, who had traveled all the way from the continent of Africa just to meet her boss.

'No they are not out of town, Dr. Foster has taken a few days leave and they are both working in her laboratory across the campus, would you like me to call them up for you'.

As she ruffled through the internal telephone directory in front of her, Joubert held up his hand in mock protest, as he pretended to consult his pocket diary.

'No that will not be necessary, I don't want to disturb them at their work...I think I have Linda's home address somewhere in my diary', Joubert purposely used Linda's Christian name as if he knew her well. This well rehearsed tactic threw the receptionist off guard and as Joubert pretended to fumble for his diary in his briefcase the girl felt obliged to assist him. Contrary to the Institute's security protocol and wanting to be helpful, she read out the address and telephone number of Linda De Vaal from the internal staff directory located in a drawer beneath her desk.

Joubert flashed her another gracious smile, the information that the foolish girl had just given him tallied with the information he had obtained earlier.

Satisfied now that the address he had was correct, Joubert bid the attractive but very gullible young receptionist farewell and strode out of the building, apparently in search of a taxi.

Once he had gone the receptionist frowned. She had just overheard Joubert say that he had no time for dinner with Dr. Levine, yet here he was planning to visit Linda De Vaal. Perhaps he prefers the company of his own South African friends for dinner, she rationalized, as she turned back to her mundane work.

Once he was outside the building Joubert was in no rush to go. He had no need to wait for the inevitable bomb blast, he was an experienced arsonist and the devices that BOSS made for him were totally reliable. He had no doubts that the magnitude of the blast from the charge he had set was more than enough to destroy the laboratory, the walk in refrigerator and all the precious samples within.

It was by now early afternoon and the timer on the bomb still had several hours to run as Joubert had cunningly given himself plenty of time to get clear of the building. He was feeling so smug and satisfied with his work that he sauntered over to a nearby cafe and ordered lunch. The salad he ordered was crisp and fresh but he felt uncomfortable sitting there in the multiracial restaurant. Back home at the Institute white professionals were still able to relax in their segregated plush facilities, whereas the black workers, who still knew their place, still sat in the canteen that had once been officially designated for them.

These lingering thoughts forced Joubert to curtail his stay in the racially neutral cafe and rather than linger over his coffee he rose from the table and went in search of a taxi. A few minutes later he was flagging down a passing cab and giving the driver directions to drop him a few streets away from his true destination, just as Devere had done earlier.

Once in the vicinity of his intended prey he strolled past the address he had been given, carefully checking the lie of the land before continuing on by. He paused at a newsagent a few hundred yards further on and bought a newspaper.

Strolling back toward Linda De Vaal's address, he crossed over the road and positioned himself in a dark doorway, almost directly across from her apartment.

Joubert carefully placed his briefcase at his feet and sank back into the dark shadows before opening up his bag and taking out the ceramic weapon. He then placed the unusual weapon inside his rolled newspaper and settled back in the shadows to await the return of one…or hopefully both of his intended victims.

Chapter 21:

Another Explosion in Atlanta.

Commandant Joubert continued to watch the apartment, unaware that Linda and Chris were already inside. They had both left work early that day, so that they could review Chris's latest findings and add further pertinent facts to their storyboard. They were so intent on this task, huddled over the kitchen table that they were oblivious to their surroundings and they were also too far away from the window to be seen by anyone watching outside.

They were also unaware that a deadly marksman was now lying in wait for them just across the road.

Neither they nor Joubert heard the colossal blast that reverberated through the Centre for Disease Control, totally demolishing Dr. Foster's lab and its contents. The large steel door of the walk in refrigerator was blown off its hinges, and the raging inferno that soon followed, fried all the contents within. The heat inside the refrigerator was so intense, each of the precious plasma samples fused with the molten glass that had moments before protected them.

As a fire truck screamed through the traffic, in the direction of the blast, the driver remarked to his mate that this was the second laboratory explosion he had heard about this week and muttered some inaudible jibe about the carelessness of scientists and their flammable reagents.

By the time the fire engine and its occupants reached the blazing scene, Dr. Levine had also reached the flaming conflagration, from the inside of he building. As he gazed in horror at the still raging fire inside the blackened laboratory, several bottles of toluene stored in the outer laboratory started to reach their flash point. Just as Levine peered through the smoke two of the bottles exploded and a jagged piece of broken glass flew through the doorway and struck him solidly on the temple. Dr. Levine fell like a stone to the floor, as the creeping flames of burning solvent reached his prostrate frame.

He was already unconscious and dying from the toxic smoke as the flames engulfed him and only a charred corpse remained when the security guard reached him. It took an hour for the blaze to be brought under control and almost another hour passed before the hysterical receptionist could pull herself together and call Chris Foster with her terrible tale of woe.

Chris listened intently as the blubbering girl relayed him the news and he had to tell her to calm down several times before he could make sense of her tearful ramblings. As he listened on the phone Chris repeated her ghastly story out loud so that Linda, who was now hanging anxiously on his arm, could also share the essential elements of what had happened.

'Huge explosion...Laboratory destroyed...Dr. Levine burned to death', it was all too awful to comprehend. As the receptionist started to calm down, Chris barked out some pointed questions, his impatience mirroring the concern he had for other members of the staff. 'Who else was in the lab, was anyone else hurt?'.

'No Sir...Dr. Levine was the only one near your laboratory at the time of the explosion', wailed the receptionist. 'But its lucky that it didn't happen a few hours earlier or the explosion would have also killed Dr. Levine's important foreign visitor', she sobbed.

'What foreign visitor?', demanded Chris, irate that some stranger had been poking around his laboratory. 'I gave no permission...who the hell do you mean?'

The receptionist sensed his sudden hostility over the telephone and instinctively rose up to defend herself.

'It was an important doctor from overseas, he was given an official guided tour...and escorted through your area by the Director himself', she responded.

'What was this visitors name, and where exactly was he from', said Chris now bristling with indignation.

Linda's eyes widened as Chris repeated aloud the words that he was now hearing over the telephone, 'Commandant Joubert,...from the South African Institute of Medical Research'.

Chris slammed down the phone and hugged Linda tightly for the name he had just repeated had left her pale with shock. As she clung to him protectively, the shadows outside were deepening and the sun had started to go down as Joubert shuffled impatiently in the dim doorway across the street.

He had no idea that at that moment his name was being spoken in the apartment he was watching, he was also still unaware that the two people he had come so far to kill were already sitting inside the apartment.

Joubert cursed himself for not bringing along a second bomb; he could have broken into Linda De Vaal's apartment earlier and set up a booby trap for them.

He could have rigged it to explode when the door was opened, or wired it to the light switch...or to the television. Joubert found solace in his rationalization that he had no way of knowing if both of his targets would return together and that his current plan was the better one. Too late now, he thought, he was committed to his current course of action, now he could only watch and wait.

As he gazed up at the window opposite, he detected a slight movement at the curtains. Then, as if to confirm his initial suspicions, the curtains were suddenly closed and the lights inside the apartment went on. Joubert reached hurriedly for his gun but the shadow behind the closed drapes moved too quickly from the window for him to loose off an accurate shot. He knew it would have been a futile attempt, especially at this range and with this particular gun.

Joubert snarled, 'Why don't these bloody Americans work regular hours, like we do in South Africa'.

It had not occurred to him that his prey might come home early, now he had to devise a new plan to pry his quarry from their secure little haven.

Meanwhile, inside the apartment Chris and Linda were also hatching a plan of their own. Now that they realized that Joubert was in town, they knew they were both in great danger.

They had figured out that the mutated HIV virus was the work of Dr. Joubert's brilliant but demented mind, but they still didn't know what he proposed to do with it. They also knew that somehow he had to be stopped.

Chris was first to speak.

'It's obvious that Joubert is covering his tracks, anyone that has handled the blood sample that came from Dr. Gorrie's dead laboratory worker has been eliminated'

'Except for you and me', Linda croaked, her grip tightening on Chris's arm as she continued to speak 'Oh Chris, do you think he is lurking nearby, waiting to kill us both?'.

'No, of course not Linda, I'm sure he has already left Atlanta and is now on his way back to South Africa', Chris lied. Chris knew that Joubert would not give up easily, and he knew that they were the only two people still left alive that had evidence that could expose him.

As an afterthought Chris picked up the phone and called the Centre for Disease Control. On hearing the voice of the receptionist, he probed for more information. 'Where is Commandant Joubert now?'

'Oh, he left here hours ago, but I did hear him mention that he was returning to South Africa via London, sometime this evening…I think', the receptionist replied craftily. 'Oh and Dr. Foster…I almost forgot, on his way out he asked me to confirm the address of your fiancée, he said he knew her from South Africa and he was planning to call on her today'.

Chris put down the phone again, his face a grim mask. He was not concerned that the office gossip had already picked up on his recent engagement to Linda but he was concerned at what the receptionist had said regarding Joubert's plans. He knew that she was in the habit of listening in on other people's private conversations but this time her nasty habit had proved useful.

Although Chris was deeply disturbed that Joubert knew their address, he did not voice his fears to Linda, he knew that she was scared enough already, but she sensed his concern.

'What is it Chris, what is it'.

'This man Joubert is a very disturbed and dangerous animal, but I'm not going to wait around until he tracks us down, I'd much rather that we go after him instead'.

'Chris, that's too dangerous, we must call the police'.

'No Linda, we don't have enough evidence and a police investigation would take far too long, Joubert has got to be stopped…and he has to be stopped now'.

Linda knew better than to argue with Chris when he was in such a determined mood and she also felt somehow relieved that her fiancée was taking the offensive. As if to stiffen his resolve Chris continued.

'No more waiting around for him to find us, we'll be the stalkers not the prey. Lets work out a plan to track this murderer down and put a stop to his horrific experiments. If that means following him to London, or even all the way back to his lair in Johannesburg, then so be it'

'A trip to London Chris, in that case I had better call my Uncle Alan, I know he will help us,' said Linda, visibly brightening at the thought of seeing her Uncle.

'Who the hell is Uncle Alan, I don't recall you ever mentioning him to me before Linda?'.

'He's my late Father's younger brother. He left Cape Town when he was a teenager and joined the Royal Air Force where he has spent the last twenty-five years, working in bomb disposal I believe.

'If he knows anything about bombs and weapons then he's certainly the man we need Linda, how come you have never mentioned him before?'.

'Oh I only hear from him occasionally, he's usually off fighting in some war...Iraq...Falklands or somewhere, but I'm his favorite niece, he's always there when I need him'

'Listen Linda, I need to nip out for a few things, so why don't you phone him now and I'll be back shortly with supplies that we are going to need for our trip'.

'Please be careful darling Chris, who knows where Joubert may be lurking', said Linda, not realizing just how close their determined pursuer was.

Chris hugged Linda briefly, grabbed his wallet and headed out the door before Linda had time to protest. As Chris burst through the front door, his sudden appearance on the street caught Joubert completely off guard. Up to that point he had been focusing all his attention on the upstairs window of Linda De Vaal's apartment. As Chris disappeared rapidly down the street Joubert knew that he had missed the moment.

Unable to reach his concealed weapon fast enough to get off an accurate shot he had stamped his feet in annoyance. Joubert was normally a deadly shot, in fact he could shoot the eye out of a rabbit.

A remarkable feat that he had often demonstrated to his troops on the warren riddled shooting range back home.

But then he had been using his trusted Berretta automatic, not the pathetic weapon he had now. Joubert's thoughts continued their frustrated ramblings. The ceramic weapon in his hand might be great for fooling airport security but it did not have the accuracy of his own customized pistol back home. Nor did it fit snugly into his quick release shoulder holster, a cunningly made device that always gave him an unfair advantage whenever he absolutely needed it.

Joubert had also faltered briefly when Chris had first emerged from the doorway, in case it was the wrong man, but now he was sure. As Chris had run by, Joubert had recognized him from a photograph that he had seen in Chris's office. Joubert settled back and waited for Foster to return. This time he was determined to be ready but he still had a few moments to rehearse the rest of a plan running through his mind.

Once he had disposed of Foster he would rush up the stairs, to Linda De Vaal's apartment, break down the door and give her a taste of his particular brand of lead poisoning. By the time the police arrived he would be long gone. As he reflected on this last thought he sank back again into the shadows and awaited the return of Chris Foster.

Forty minutes later Chris came jogging back down the street. He was carrying a small bag of supplies and as he drew closer Joubert took the ceramic weapon from inside the rolled newspaper and gripped it in his right hand. There was no safety catch on the gun and it felt bulky in his hand.

As Chris drew almost opposite where Joubert lay concealed, the killer rested the pistol on his left forearm and took careful aim at Chris's sweaty temple. Then just as Joubert squeezed the trigger, a taxi passed between them The gun had no silencer and exploded loudly before the speeding bullet shattered the side window of the cab as it went speeding by. Chris heard both the noise of the gun and the shattering of glass as the lethal projectile impacted the taxi window and spewed pieces of safety glass onto the street.

Chris looked in Joubert's direction as the. thwarted killer shrank back into the shadows and loosed off another shot but his haste for concealment skewed the accuracy of his second shot. The bullet missed Chris by inches and the brickwork above his head spat out shards of broken masonry that showered down on Chris head like concrete hailstones.

By the time Joubert was ready for his third shot Chris's fast moving frame was behind the heavy wooden front door and he was leaping up the stairs, two at a time. Chris burst through the door and Linda flew to his side. She had heard the commotion outside and she had already noticed the trickle of blood, caused by a sharp piece of splintered brickwork that had struck him a glancing blow, trickling down his face.

'Who was it Chris...was it Commandant Joubert?'.

'Yes, I recognized him right away...from the photo, that I printed off from that article on your computer'.

As Linda fussed over him Chris assured her that it was just a flesh wound and tried to make light of both the injury and the situation. 'There is a doctor in the house, in fact there are two', he joked, forcing a smile to Linda's trembling lips.

As soon as she was satisfied that he was all right Chris pushed her away gently and instructed her to turn out the lights and keep well away from the window. Down below Joubert saw the lights go out and realized that his cover had been blown so he prepared to leave.

If Foster called the police he did not want to be around when they arrived. He had now lost the element of surprise, so it made no sense to go charging up to the apartment, the noisy gun had already attracted far too much attention. The taxi he had inadvertently shot at had not stopped but all around him windows were starting to open and people in the vicinity were shouting out their outraged protests at the noise.

Throwing the weapon down on the ground in disgust Joubert sprang out of the shadows and sprinted off down the street. Within minutes he was well clear of the area and as he slowed down to a walk, he hailed a passing taxi. Jumping inside he gave the driver terse instructions to head for the airport and then settled down in the dark back seat.

Once he was safely on his way he remonstrated himself for his pathetic performance. Fortunately none of his underlings had witnessed his botch up and the loss of his gun and briefcase were of no real consequence. The briefcase was empty; Joubert had thoughtfully removed all his notes and identification tags during lunch at the restaurant. As for the ceramic gun, it was useless and he was glad to be rid of it.

Joubert's only concern now was that his cover was blown and that Foster might be able to recognize him. Joubert considered his options.

Safer to return to South Africa and then send another agent, one that no one would recognize…one of his other hired killers could take care of them now, Joubert sneered. Joubert had left his overnight bag at the airport but the ever cautious Commandant had made sure that it contained no incriminating evidence, just some dirty socks and underwear, he thought to himself.

Meanwhile Linda had staunched the blood from Chris's head wound and they were both now sitting in the darkness waiting until they were sure that the coast was clear. Chris rose from the couch first and peered through the window. Looking across the street he watched as several cars went by, their headlights illuminating the dark abyss where Joubert had laid in wait for him. Chris saw the discarded briefcase and a plastic object lying nearby, but he still waited another ten minutes before venturing outside and crossing the road to retrieve Joubert's discarded stuff.

To his disgust the shabby leather briefcase was empty and the strange gun lying there gave up no clues regarding its previous user. Chris had no use for guns but he did not want it to fall into the wrong hands so he gingerly placed it inside the briefcase and took them back to Linda's apartment.

'Joubert has obviously gone, he left his gun and his empty briefcase…now it is our turn to go after him'.

'Chris, in all the excitement I forgot to tell you what I did when you where out picking up our supplies'.

'Tell me as we are packing Linda, I don't want to lose any more time getting after that madman'.

Linda was surprised at his determination but she dutifully started throwing clothes into a suitcase as she told what had happened during the time he was out. Chris listened attentively as she told him that she had contacted her Uncle in England, who had promised to meet them on arrival in London.

She went on to tell Chris proudly that she had also booked them two tickets on a flight from Atlanta to London with an ongoing connection to Johannesburg. 'More on our airline reservations later', she smiled somewhat mysteriously.

'Please Linda, don't tell me you've spent all our hard earned savings on First Class tickets', he joked.'

I'll tell you later Chris…when we get to the airport'.

'Don't forget to pack our passports', called out Chris.

Don't worry I've already taken care of that', she replied smiling to herself and crossing her fingers behind her back. It was her way of hoping that Chris would not be too cross when he found out about the passports and the airline reservations she had made.

'Chris, are you absolutely sure you that you want to go through with this very dangerous and foolhardy plan?'.

'Damn right I do, no assassin is going to shoot at me like some wild animal and I'll tell you something else Linda. This maniac Joubert has somehow engineered a deadly AIDS virus…now both he and his vile creation must be destroyed'.

'You are right Chris, millions of lives could depend on us, if we don't stop him…then who else will?'

She left her last question unanswered as Chris carried their suitcase downstairs and she telephoned for a taxi.

Ten minutes later they were both sitting in the cab, heading for the airport, hot on the trail of Joubert.

As Chris put his arm around her in the taxi, Linda again crossed her fingers behind her back.

Chapter 22:

Hunter and Hunted fly into London

The Atlanta airport terminal was busy and Chris was about to cut through the crowd and head to the check in desk when Linda suddenly pulled him to one side. 'Don't be cross Chris, but as an added precaution I made both of our reservations under a fictitious name.

'How the hell did you do that,' said Chris, more shocked with her ingenuity than by her deception.

'You remember Wendy Hughes…The friend of mine who works for Roche Diagnostics…The one who is such an expert with computers…well I called her up when you were out getting supplies…She used her terminal at work to input our false names directly into the airline computer.'

'So who are we pretending to be Linda? And how the heck do we get through immigration and customs?'.

'I thought of that too Chris, we are now a respectable married couple…Mr. and Mrs. De Vaal. I brought along my Father's old passport for you to use and I am going to travel on my old South African passport'.

'Very devious Mrs. De Vaal but what if the authorities spot the difference between me and your Father's photo…And of course the wrong date of birth?'.

'Don't be such a worrier Chris, you do look like my Father when he was younger, and they will be far too busy at customs to check everyone's date of birth'.

'You are a crafty one Linda but I see the wisdom of travelling under a different name, especially after what you have told me about Joubert and his BOSS cronies. If Joubert suspects he is being followed he will probably be on the lookout for me as Foster, I doubt if he will guess that we are travelling together under your surname Linda.'

'Exactly, and when I told my Uncle Alan, he also thought it was a good idea, especially when I explained the reason for our mission and the danger we were in'.

As Chris and Linda were reviewing their plan to sneak through security at Atlanta airport as Mr. and Mrs. De Vaal, Commandant Joubert was already well on his way to London. His connecting flight from New York had left on time but as he sat in comfort, in the first class cabin of the British Airways jumbo jet, he was not a happy man. He was still cursing his aborted effort to eliminate his two adversaries and he mouthed the names of Foster and De Vaal under his breath.

Even the expensive champagne, that the attractive young stewardess had just served him, failed to brighten his mood. As she passed down the first class cabin delivering her bubbly welcome to the rest of the privileged passengers, Joubert stared at her lasciviously.

His eyes were drawn to her shapely hips...hips that were now swaying provocatively beneath her thin blue skirt. As the tall leggy blonde stretched from the aisle, to reach one of her window-seated passengers her white blouse tightened across her firm young breasts. Joubert licked his lips and rubbed his groin longingly.

Several days had already passed since his visit to Madame Lazelia and his encounter with the black dancer at Sun City had been a sexual disaster. As he continued stroking his loins his thoughts fluttered between the failed mission behind him and the rising need for sexual gratification that now gripped him.

He had no doubts that his cover had been blown and it had been smart to leave Atlanta before the authorities picked him up and started asking awkward questions. Perhaps Dr. Foster would try and track him down, he smiled at this last thought. Joubert would be more than happy to confront the foolish young doctor, especially back home in South Africa, where his resources and status would give him a clear advantage.

The thought of tackling Foster on his own turf moved up his dark mood a notch as he mentally formulated a new plan. He would make a stop off at Heathrow airport in London and contact one of his agents at the South African embassy. He could then use the embassy computer to monitor all passengers flying into London and into South Africa. If Foster was foolish enough to follow him back to his lair, Joubert would know about it and be waiting for him.

The Commandant smirked, a stop off in London would also give him an excuse for a side trip. There he could sample the delights of one of his favorite prostitutes, one he had enjoyed before in the notorious red light district of Soho. Joubert had not been there for many years but he had no doubts that the menu the whore offered now was just as varied and depraved as he happily remembered it.

This last thought eased his earlier self-recriminations as he settled back in his plush leather seat and prepared to sleep away the next few boring hours.

Down on the ground Chris and Linda were through security at Atlanta airport and were now heading nervously toward the customs and immigration check point. They had experienced no trouble picking up their tickets at the busy check in counter. The clerk there had been far too rushed to bother checking their tickets and passports. Now they had to get through the customs check point.

As the queue ahead of them edged closer to the glass booth, the blue uniformed officer ahead was casting his eyes closely over each passenger as he matched their faces with their passport photograph in front of him. Chris felt his mouth go dry. Linda was also nervous as she grasped his hand, her fingers tightening around his.

At the designated moment they both stepped forward together, and stood facing the young customs officer across his protective counter. The newly-qualified young man was too busy ogling at Linda, her seductive smile warming his drab little kiosk, to spare more than a cursory glance at their tickets and their passports.

Lucky guy thought the customs officer…The young man now before him had every reason to be happy…such a gorgeous looking woman hanging on his arm. The customs officer's eyes continue to traverse Linda's shapely body and then headed down to where Linda had cunningly intended them to go. At that same time Chris also saw that Linda had left the top two buttons of her blouse undone.

The customs officers gaze was now fixed on the deep mysterious cleavage between her up thrust breasts, as he distractedly rubber stamped both their documents. Chris and Linda breathed as sigh of relief as they stepped beyond the checkpoint and headed for the departure lounge.

'Pretty crafty Linda, that poor young lad almost dived head first down the front of your blouse, it's a wonder his rubber stamp didn't fall out of his shaky fingers'.

'Oh hush Chris, at least we got through customs and immigration without any hassles', Linda blushed.

At the departure gate they boarded the waiting plane and as they passed through the first class cabin they both gazed enviously at their plush surroundings. As they continued on through to the cramped tourist section, Chris found their seats and then stowed their suitcase in the overhead rack. As Chris sat down he gawked enviously at the passengers who were now seated in the first class cabin and who were now sipping their champagne. He had no idea that in a similar cabin, high in the stratosphere above the Atlantic Ocean, Commandant Joubert was also sitting in First Class and drinking champagne.

Once Joubert's glass was empty, he pressed the button of the seat recliner and settled back to sleep. His slumber continued uninterrupted until the sound of the heavy landing gear, locking into place, woke him up. Rising from his seat Joubert paid a quick visit to the cramped little washroom, where he splashed some Balenciaga after-shave on his face. The sweet smelling perfume, thoughtfully provided for the first class passengers, was not to his liking, but its astringent properties refreshed his travel weary skin.

Returning to his seat he sensed that the big jumbo jet was now starting its final approach into Heathrow airport. The international flight must have had priority for it was soon down on the ground and parked at the terminal, ready to disgorge its four hundred passengers. The first class travelers were the first off the plane and Joubert's long strides took him ahead of even the passengers that had been seated in front of him. He immediately sought out the nearest telephone.

Selecting one that offered the most privacy he dialed a special number at the South African embassy in London, a secret number he had memorized many years before. The agent at the other end of the line almost clicked his heels together deferentially as he recognized the voice of his superior officer and automatically sprang to attention.

Without wasting time on pleasantries Joubert spat out his specific instructions. Satisfied that his agent would now monitor all U. S. flights coming into any of the airports in England and all flights going directly to South Africa, he put down the phone. Joubert was still not sure if Dr. Foster would follow him. He also doubted that Foster would be cunning enough to fly to South Africa via Frankfurt, Paris or one of the other European connections, but thought it prudent to check.

His London based agent would use the powerful BOSS computer, secreted inside the embassy, to scan the passenger lists of each and every plane travelling to South Africa. If Foster was on any of them, Joubert would know about it. Joubert would then have some of his more brutish subordinates waiting to meet Foster. They could first work him over before finally bringing him to Joubert for further torture and destruction.

Joubert grinned as he visualized Foster screaming for mercy and spilling his guts before he was eventually executed. He was starting to dribble with excitement, beads of spittle glistening on his cruel lips as he dwelt on this thought until he reached a taxi stand outside the terminal. Despite the long line of waiting cabs, there were none being manned by white drivers, so reluctantly he signaled the first one in the line.

As he jumped in behind the driver his olfactory senses immediately protested. The strong smell of curry together with other unknown, but equally offensive spices, almost made him gag and he had a sudden impulse to leap back out of the cab. However he was anxious to get to his destination so he quickly wound down the window and told the Sikh driver where he wanted to go. The East Indian turned to leer at the mention of his Soho destination but the smell of his regurgitated breath brought only a deep scowl of response from Joubert.

Joubert had no wish to engage in conversation with this smelly peasant so he growled out his directions again and curtly told him to 'Step on it'

The driver took off and glanced only briefly in his rear view mirror, he instinctively knew that the powerfully built passenger in his back seat was not to be messed with. The sooner he dropped this ominous looking stranger at his sleazy destination, the better.

On arrival at the now darkening streets of Soho, Joubert tapped the glass divider when he spotted a familiar landmark in Wardour Street and the cabby screeched to a halt. The driver, a refugee from Afghanistan, did not protest when Joubert only gave him the exact fare that had been shown on the meter. The mean glint in his passenger's eyes reminded him of the Mujahadeen warriors who used to descend from the mountains to pillage his little village back home. He had been terrified of those man then and he was scared of this man now so he averted his eyes as he snatched the money from Joubert's outstretched hand and drove away.

Joubert strolled down Wardour Street for a few hundred yards before turning into one of the many dark avenues that cut across the main thoroughfare. As he negotiated the ever-darkening side streets, the tall old buildings lining either side of the street became even more dilapidated. He ignored the occasional pimp and assorted human flotsam that he encountered along the way and they in turn wisely avoided him, but a drunk who was sprawled in a doorway soon had reason to regret his request for money, receiving for his trouble a well-aimed kick. The drunk had rolled back in agony, moaning and groaning as the bright blood from his broken nose mingled with the dried vomit on his ragged jacket.

As Joubert moved deeper into the catacombs of Soho he noted that the street lamps were getting less frequent. They also shone less brightly and the light from their grimy lenses seemed unable to penetrate the deepening gloom. Suddenly their dim glow was surpassed by little oblongs of bright light, some in columns of six or eight that were now winking at each and every doorway.

On closer inspection a passing stranger would have seen that they were the illuminated names and bell pushes of the occupants that resided within.

A more naive stranger would have been surprised to read the names and captions lit by these tell tale billboards…Denise… Monique…Miss Leather. Some went on to offer lurid descriptions of what each lady inside had to offer but it was one that promised the most erotic of pleasures that Joubert looked for.

After casting his eyes over more than a dozen of them he finally found one he remembered from years ago…Hitler's mistress…Extreme domination and delight…No fantasy or perversion too extreme for this nazi of the nineties…Joubert stroked his fingers lovingly over the faded yellow card, his mind already savoring the devious sport awaiting him inside as he pressed the lighted bell push.

He remembered the code from his last visit. If the bell was lit up, the lady inside was ready and waiting…If the light was out, some other customer was enjoying her favors. Climbing the dark stairs inside, Joubert reached the dimly lit landing where a scantily clad women in leather jackboots met him. On her head was a nazi officers peaked cap, its severe shape in stark contrast to her frizzy white hair. As she pulled him into her dimly lit apartment he could see that she was wearing a black suspender belt over her leather panties and her neck was girded with a spiked leather collar. Her wrists too were encircled with spiked leather bracelets, their sharp steel points glinting in the gloom.

Joubert turned to leave, instead of finding her attractive her bizarre appearance made his stomach convulse. Under the nazi paraphernalia her bleached blonde hair and her pale sallow skin was coarse and uninviting. Even her breasts, which were fully exposed, were small and wizened and Joubert was repulsed at the thought of touching these varicosed bags of skin…Hitler's mistress indeed, she looked old enough to be Hitler's Mother. This was not at all like a much younger Joubert had remembered her; the passage of time and the ravages of her profession had indeed taken its toll.

The old crone sensing that Joubert was about to leave took a whip from her belt and snaked the leather thong around his neck, reeling him in as she attempted to draw his lips closer to hers. Joubert's mind and patience snapped. He had come here to satisfy his lust but now his thoughts turned to death and destruction. He jerked the old crone towards him with such force that she cannoned off his brawny chest. Then as she rebounded back again, he gripped her scrawny neck in a vice like grip.

As if killing a chicken Joubert twisted his wrist in a way that he had been taught during unarmed combat, and the harlot danced helplessly at the end of his arm. Her scrawny neck could not tolerate the traction that his powerful grip was imposing so it snapped with a loud crack. For a few moments he held the dead whores body, dangling uselessly from his outstretched arm, and then with a jerk he flung it down the stairs.

He was now sweating profusely, not from exertion, but from the excitement of the kill. He had foregone any pleasures of the flesh that this pathetic creature might have given him but somehow her death had left him sexually satiated. How very interesting, thought Joubert, as he stepped over the dead body and headed out the door of the dank building, killing her had been better than having sex.

The feeling of euphoria now coursing through his demented brain was so satisfying that he decided to forgo further incursions into this weird wonderland.

As he made his way to the main thoroughfare he was only once more accosted on his way back to a taxi. This time by a voluptuous black woman, who looked like she had the equipment she needed to meet his desires. As she called out in a refined English accent, 'Looking for a nice time Darling,' her accent left him puzzled and un-aroused.

England had indeed sadly deteriorated, he thought, they even let their black servants imitate the posh British accents of their once great aristocracy.

Had a passing stranger been able to read his innermost thoughts they would have pitied this bright intelligent man, whose intellect was unable to rise above his inbred racism. Even a child could have told him that these black skinned people had been born and raised in England and were more 'British' than Joubert could ever hope to be.

Reaching the bright lights of Wardour Street, he hailed a passing taxi and gave the driver instructions to take him back to Heathrow airport. He had no desire to stay any longer in London and he was hoping he could catch an earlier flight back home to South Africa.

As his cab negotiated the airport terminal, heading toward the departures area, an incoming jumbo jet carrying Chris and Linda was circling overhead. Had Joubert known this he would have somehow reached up and tried to pluck them from the sky.

By the time Chris and Linda's plane had discharged all its passengers into the terminal, Joubert was rapidly making his way to the South African airways ticket counter. He had no way of knowing that only the concrete floor beneath his feet separated him from the ever threatening couple that were now pursuing him.

Chris and Linda were likewise unaware that only the ceiling above them shielded them from an archenemy. As Chris and Linda moved into the customs and immigration area their thoughts again focused on the risk of being caught with their fraudulent papers. Suddenly Linda's face broke into a grin and she opened up her arms to embrace the tall imposing figure now striding purposefully toward them.

'Uncle Alan, how wonderful to see you.'

'Welcome to England, favorite niece of mine, I assume this young man hanging on your arm is your fiancé.'

As the stranger disengaged himself from Linda's arms, he extended his hand toward Chris in warm welcome.

Chris gripped the outstretched palm and gazed into the clear blue eyes of Alan Cobb. They were dancing with merriment now but they also harbored a piercing look of steely resolve. Well over six feet tall, Alan Cobb was a lean but powerfully built man with just a few streaks of gray in his otherwise jet black hair. The deep lines on his craggy face a testament to the pressures of his job over the past twenty-five years, not his age.

Chris had learned on the plane that Alan Cobb had joined the Royal Air Force as an armament apprentice at the age of seventeen and had completed his initial training at RAF Halton in Buckinghamshire.

He had served in Germany and Suffolk before his superiors had recognized his talent and induced him to apply for officer selection. Following his advanced technical training he had graduated as a Flying Officer and had served in Iraq, the Falkland Islands and Northern Ireland. as head of a bomb disposal squad.

Prior to his recent retirement he had risen to the rank of Wing Commander and had been in charge of all bomb disposal teams in the United Kingdom. Such was his skill and knowledge; he had been brought out of retirement and was now working as a civilian, on various covert operations. Operations that were known only to the most senior personnel of the armed forces and the government.

Linda had also told Chris on the plane, almost in a whisper, that Uncle Alan was on the 'IRA hit list' because of his close connections with the police and the armed forces anti-terrorist units.

Chris had been impressed by Linda's description of her Uncle and now that he had met him he was even more impressed, he also felt immensely relieved to have him along to help. Chris was still to find out just how resourceful and helpful Alan Cobb was going to be. Chris's thoughts were interrupted as Linda spoke.

'How did you get inside the customs lounge Uncle, I didn't expect to see you until we got through security.'

'Oh I have my connections', laughed Alan, 'I also wanted to ensure you didn't have any hassles getting through our British customs with your bogus papers'. With that Alan grabbed them both by the arm and ushered them through a side door away from the long line of newly disembarked passengers.

As they passed through the security door marked 'No admittance, authorized personnel only', a uniformed man inside, rose to his feet and snapped off a salute. He had obviously recognized Alan Cobb right away. In response to Linda's raised eyebrows Alan chuckled as they all strode past the security guard.

'Yes I've done a few favors over the years for both the police and the Special Branch, it certainly eliminates waiting in line at the airport', he smiled.

Once outside the airport Alan ushered them to his car. His sleek silver Jaguar was parked in a spot marked 'No parking, official airport use only' and Chris knew that Alan had connections that would come in handy. As Chris and Linda were driven out of the airport, most of their fellow plane passengers were still shuffling their way through the slow moving customs queue, and some of them were fuming at the delay.

Commandant Joubert was also fuming, but for a different reason.

Chapter 23:

The Hawk Spots His Prey.

The arrival and sudden departure of Chris Foster and Linda De Vaal from Heathrow had not gone unnoticed. Their appearance in the Customs and Immigration area and their rendezvous with the tall dark stranger had also been seen from above. The way that the stranger had ushered both of them away through a side door, would have aroused the suspicions of any astute observer…and it had.

Commandant Joubert had been walking along a balcony overlooking the arrival area, when he had glanced down below and seen a face he instantly recognized. He could not believe his eyes, there…down below…No it couldn't be…indeed it was…It was the same man that had evaded him in Atlanta…Christian Foster!.

Joubert almost spat out the venom of his anger as he focused on the three figures below him. The woman hanging onto Chris's arm must be Linda De Vaal he surmised as he burned this new face into his memory. He had the distinct feeling that he had seen her before, not recently, but somewhere in one of his old memory banks, her face already occupied a minor slot.

He knew that she was a virologist from South Africa; perhaps she had been one of his students, one of many eager faces at one of his lectures back home.

As these thoughts raced through his mind at breakneck speed he also concentrated his attention on the tall dark stranger who had also just joined them.

He was obviously someone with influence here at the airport, the way he had spirited Foster and De Vaal through a security exit. None of them had waited in line for the obligatory clearance through Customs and Immigration and this made Joubert very wary of this new man on the team.

The Commandant had been so angry when he recognized Foster, his first reaction had been to reach for his gun, but then he remembered that he had discarded it in Atlanta. It was fortunate; it would not have been a smart move. Heathrow airport was bristling with police and security personnel; this was not the place to carry out an assassination and then hope to evade capture. Only the Islamic suicide bombers were fanatical enough to attempt an act of violence in this terrorist sensitized airport.

Joubert was a ruthless killer…but he was not suicidal. The Commandant smirked, at least he knew that Foster and De Vaal were following him and as they had not seen him watching them from above, he had edge. Now he would continue on home and prepare a hot reception for their arrival in South Africa. The thought of all the agents he could muster back home made him realize that his prey had somehow arrived in London without his knowledge, despite his specific orders.

Angry again, Joubert strode to the nearest phone and vented his spite on the agent who answered his call. 'What the hell went wrong…The person you were supposed to be monitoring…Foster, just arrived at Heathrow…Why was I not informed of this earlier?.'

Before the bewildered agent had time to splutter out a reply, Joubert ranted on again.

'You will immediately double check your computers…I will call back in exactly five minutes, if you do not have an explanation you will be severely disciplined.' Joubert slammed down the phone and stood drumming his fingers on the side of the phone booth as he watched the fingers on his watch measure out the passing time. Exactly five minutes later he picked up the phone receiver and called the embassy again.

'So what excuse do you have…A computer malfunction?' Joubert sneered sarcastically.

'No Sir, the computer is working perfectly, I have double checked our system and no one named Foster was travelling on the flight from New York that has just arrived at Heathrow.'

Before Joubert had time to reply the agent continued…'Only two people departed from Atlanta and then made the connecting flight from New York to Heathrow, and they were listed as a Mr. and Mrs. De Vaal, the agent concluded smugly.

Joubert immediately knew what had happened and he curtly snapped out his next command without even bothering to apologize to his falsely accused agent. 'You will immediately modify your computer search and ensure that it now covers anyone and everyone travelling under the name of Foster…Or De Vaal.

Slamming down the phone without waiting for a reply Joubert was satisfied that the BOSS surveillance system was functioning properly in London. Unauthorized access to airline passenger manifests was illegal but counter intelligence networks and powerful computer search engines were standard equipment at all their international BOSS outposts.

It was Joubert's error that they had only been searching for Fosters name, but he certainly wasn't going to admit that to one of his underling at the embassy. Commandant Joubert had underestimated the cunning of his two pursuers but he would not make that mistake again. As Joubert headed for his plane he stopped to pick up a copy of the Daily Express at a news stand before he boarded the Boeing 747 that was waiting for him on the tarmac.

Once aboard and settled in his preferred first class cabin he settled back in his seat and opened up the newspaper. Flicking through the pages of the English tabloid his eyes fell on an article about shocking crime rates being recorded in the city of Johannesburg. He threw the newspaper to one side in disgust.

He had only been out of the country for a few days and already his once great country was hitting the international headlines with tales of growing turmoil and moral decay. The sooner he got home and put into motion the start of his perverse panacea...the better.

Chapter 24:

Layover in London

As Commandant Joubert slept in his plush leather seat, high in the sky over the English Channel, his enemies were down on the ground in equally opulent surroundings. Chris Foster and Linda De Vaal were too surrounded by leather, in the back of Alan Cobb's sleek Jaguar XJS, speeding along the M4 motorway.

As the car turned off the highway and headed in the direction of Slough, Linda stopped her idle chatter and asked a pointed question.

'Why are we heading for Slough, I thought you lived in Norfolk Uncle Alan?'.

'My wife and family are safely based there but I have many aliases and a number of safe addresses, so that I can keep one jump ahead of all my enemies. The mad bombers in the terrorist wing of the Irish Republican Army have not forgiven me for thwarting their plans.'

'But why Slough?' questioned Linda.

'Oh I have a safe house here where I can leave you both for an hour, while I make a side trip to Hendon'.

'Hendon, isn't that where they hold the Royal Regatta,' said Chris, now joining in the conversation.

Alan Cobb guffawed loudly. 'No that is Henley Chris, Hendon is the home of our police college and it is also the location my own private armory'.

Linda shrank back in her leather seat at the mention of this last word but she reconciled herself to the fact that she now had two strong men to protect her.

'I want to pick up a few special supplies for our trip, from what you tell me about this maniac Joubert I know I am going to need them,' Alan Cobb went on.

Our trip?' shouted Linda, 'Are you also going with us to South Africa ', Uncle Alan?'

'Of course, I wouldn't miss this adventure for the world, you are going to need both me and my special equipment, if we are going to stop this madman and eliminate his deadly virus'.

Linda clapped with glee at her Uncle's last statement. They continued on in silence, each of them thinking about the unknown dangers that lay ahead of them, until the car tires crackled as the Jaguar swung into a gravel driveway. Alan parked the car and led them into a large two-story farmhouse that was set back from the main road.

The house had been in darkness as they arrived and as Alan led them inside the main door he clicked on the lights in each room and showed both of them around. The house was sparsely furnished but it had two bedrooms and a well-equipped kitchen, with plenty of food in the refrigerator.

'What! no woman living here?,' teased Linda.

'Don't be cheeky Linda, I am a happily married man and this is a safe haven for me whenever I need it'.

'Who keeps it stocked up with food,' queried Chris.

'My friends at the Defense Department now stop asking silly questions and go and get some rest'

Alan ushered them both up to the main bedroom, carrying their suitcase as he led them the way. 'There's two single beds in here, you can use them both or just one, depending on how tired you both are'. As Alan threw out this last comment his eyes twinkled. He had taken an instant liking to Chris Foster and he knew that Linda would be safe with him during the time he was away in Hendon getting his supplies.

Despite their cozy surroundings and her Uncle's lecherous suggestion, they were both very tired, so after undressing, they both fell asleep innocently in each other's arms. Several hours later they were up, dressed and showered before they heard the crunch of the Jaguar's tires as it pulled into the gravel driveway. Alan Cobb came bounding in through the doorway, obviously happy with the success of his little side trip.

He placed a small suitcase on the kitchen table and as he opened it up Linda let out a gasp of mild surprise. It contained a strange looking gun and several aluminum tubes, which looked like insulin injectors. 'I had no idea you were a diabetic,' said Linda, remembering her medical training.

'I'm not, but I want the security people in South Africa to think I am, I don't want them examining these little babies too closely'.

Alan Cobb went on to explain that he had no trouble getting through security in countries where the British Government had jurisdiction, but as South Africa was beyond their influence he needed to take precautions. As he laid out each of the items from his little suitcase he explained their deadly purpose to both of them. 'The 'insulin injectors' are high explosive incendiary devices and what appears to be a 'blood glucose monitor', is an electronic timer for the explosives'.

'Damn cunning,' exclaimed Chris, 'but those little silver tubes don't look big enough to do much damage'

'Don't underestimate them Chris, they contain a newly developed high explosive, far more powerful than plastic explosive or Semtex. Each of these little babies is capable of completely destroying an armored tank'.

'What about that gun, it's not like any that I have ever seen before,' chipped in Linda, anxious to contribute.

'This dear niece is the British Army's latest discovery. What you see in front of you is a weapon developed by our Belgian allies, a gun so powerful that the bullet it fires can penetrate Kevlar body Armour. A pal of mine at the ministry has allowed me to do some field trials on this new weapon...and the site I have chosen is Johannesburg.'

Linda and Chris were wary of guns but they respected Alan Cobb's expertise in weaponry and were grateful that he would be along and knew how to use them.

'Now children I have a little something for you both ', he said, pushing a fat yellow envelope across the table.

On opening it up Linda found two British passports, and although neither of them bore their real names, they each carried a recent picture of them both.

'How on earth did you get these Uncle Alan...both these photographs of us were taken quite recently'?

Alan went on to explain that during the time he had been at Hendon, his friends at Interpol had accessed the motor vehicle department computer in Atlanta. They had transmitted a digitized picture from Linda and Chris's driving licenses and wired them to Alan. Yet another of Alan's 'friends' had taken the transmitted images and affixed them to two genuine British passports.

'You are clever Uncle Alan, so now we are going to travel as Steve and Anne Leonard, however did you come up with those names,' Linda teased.

'Oh those are two good friends of mine in Canada, people who I know will not be travelling to South Africa over the next few weeks.'

'That's just great!, I always wanted a second passport, especially a British one,' said Chris excitedly.

Don't get too excited Chris, its only a temporary loan, they both have to be returned to me when we get back from our mission…safely'…Alan added confidently.'

Now if you two have finished ogling at your new passports we should pack our things and get ready to leave. The flight I booked us for Johannesburg leaves in three hours and despite my connections it will not wait beyond its departure time for Mr. and Mrs. Leonard, or myself.'

'Yes but knowing your contacts Uncle Alan, we will have no trouble finding a parking space, nor will we be delayed by customs and immigration,' joked Linda.

Chapter 25:

The Hawk flies back to his roost

Commandant Joubert peered out of his window in the plane and saw the strange hills below, yellow hills that almost encircled the sprawling metropolis of Johannesburg. The huge golden mounds were the residue of millions of tons of earth that had been excavated from the ground, hundreds of feet below them, a living testament to the hordes of poor black miners who had sweated and died underground to extract the white man's precious gold. Joubert did not share these sentiments; the yellow outcrops they were approaching merely meant that he was home.

As the Boeing 747 reached the runway its multiple tires hit the tarmac simultaneously and spun in unison before the aircraft came to a controlled stop in front of the terminal. As the passengers arriving at Jan Smuts airport spilled down the gangplank and headed for the clean white building, Joubert immediately broke away from the pack. It was now his turn to bypass the security clearance that was waiting for all the other passengers. He was back in BOSS controlled territory now and his membership in this organization gave him influence at the airport. One of his agents waited nearby and fell in step beside him. Together they marched in silence through their own special exit, to the car park outside the terminal building.

The agent had dutifully brought Joubert's Mercedes and he had thoughtfully left the engine running. The air conditioner inside the car had already sweetened the hot dry African atmosphere, which had hit them like a hot wall as soon as they had left he terminal.

The Commandant did not speak as the agent opened the rear door of the limousine for his master to step inside. As the big luxury car cruised its way to the Northern suburbs, Joubert sat in the back and decided that he would not go in to work for the next few days. Not during normal working hours anyway.

There was no pressing work in his laboratory and he did not want to hear from his mentor in Pretoria as this would not be the time for explanations or lame excuses. If necessary he could visit his laboratory during the evening, when no one was around, and when all incoming calls from Pretoria would be intercepted and recorded by the emergency switchboard.

In less than half an hour his still silent driver was delivering the Commandant to the door of his house in Bryanston and was parking Joubert's car in the driveway. As the agent prepared to leave in his own vehicle, Joubert issued a solitary command. 'If word comes in regarding the arrival of Foster and De Vaal, I am to be contacted immediately...Day or night.'

The agent did not need any further explanation, he had already been briefed on his Commandant's wishes and he knew that his boss was in no mood to talk.

Joubert entered the house and went straight upstairs to his bedroom, it had been a long and tiring journey, now all he wanted to do was sleep. Joubert rose early the following morning and went outside for a swim. He was determined to shake off his jet lag and he knew that some brisk physical exercise would soon get his tired body back in shape. After swimming as many laps as his body would tolerate he climbed out of the pool, crossed the grass and stood toweling himself down in the little cabana.

The red warning light was still glowing steadily on the wall but Joubert expected this. He had not cancelled out the alarm, by pressing the little toggle switch inside his special refrigerator...there had been no need...the contents had already been violated. The glowing red light also served as a reminder to Joubert that it was not wise to keep all his eggs in one basket.

As he cast his mind back it seemed like a lifetime since the black handyman had first stolen the tube of blood, a tube so precious that Joubert and his henchman had traveled thousands of miles to cover up its loss. The fact that they had been unsuccessful still gave him cause for concern and in response to this last thought he strode to the kitchen and opened up his freezer.

The six tubes of blood he had brought home from the laboratory were still safe and sound at the back of the refrigerator. These were the second string to his bow, in case anything else happened to the ones in his laboratory.

He also had a third string to his bow, one that even his mentor in Pretoria didn't know about. Only Joubert and one other man...a loyal officer who had served faithfully at his side throughout all his campaigns, knew the whereabouts of his other mutated creations.

Joubert smirked as he recalled how he had fooled his wealthy and powerful superior in Pretoria. It had not been easy wheedling sufficient money out of him to equip not one, but two laboratories. Fortunately his mentor was not a virologist and Joubert had been able to convince him that the cost of all the special supplies he needed, was much more than it really was.

An invoice forgery here, an astute purchase there, Joubert had been able to duplicate all the laboratory supplies he needed, without anyone else knowing about it. Commandant Joubert was a fiercely loyal man, but only to himself…And to his own cause.

For the next two days Joubert concentrated on getting himself back into shape. He made only two visits to his laboratory at the Institute for Medical Research, both at nighttime. The only person that he met there was his trusted guard, and he was sworn to silence.

No one else had seen him come and go in the darkness…Or so he thought. There had been no further word regarding the whereabouts of Foster and De Vaal and Joubert was beginning to wonder if they had given up the chase in London and gone back home.

Nevertheless he still intended to implement his master plan within the next 24 hours.

Chapter 26:

One Nest is Discovered

The arrival of Chris Foster, Linda De Vaal and Alan Cobb in Johannesburg, three days earlier, had been uneventful. The plane had landed and Chris and Alan had disembarked together, each with a camera around their necks...cameras that Alan had thoughtfully purchased at the duty free store at Heathrow airport. Together they had strolled through the airport and into the immigration area as if they had been friends for many years.

The inspecting customs officer had no reason to doubt the plausible reason for their visit...they were amateur photographers going to Kruger Park to shoot big game.

The customs officer had laughed at their comment. 'As long as you shoot them on film, not with a rifle,' he had called after them.

Linda had followed them through a few minutes later. Wearing rimless glasses and with her normally long blonde hair tied up in a neat bun, she had looked every inch the schoolteacher she was pretending to be. The officer conducting her inspection had accepted her word that she was arriving in South Africa on vacation.

Once outside the airport the three bogus travelers had met up again at the Avis rental counter where the vehicle that Alan had booked for them was waiting. They had then all piled into the little white Peugeot and headed out of the city to the Northern suburbs.

It was Linda who had spotted the Balalaika hotel in Sandton and said that this was where they must stay. She liked its rustic exterior and its quaint thatched roof. Alan had agreed with her choice, he knew from the map he had brought with him that it was only about a twenty minutes drive from the South African Institute for Medical Research. It was close enough for their purpose, yet far enough away to avoid any unwanted suspicion.

Once inside the hotel they requested individual rooms, to maintain the pretence that they were not travelling together...Alan had insisted on these precautions. They then all sat together in Alan's room as he outlined his plans for their assault on the Institute. He had already unpacked his supplies and dug out a map of the Institute that he had somehow obtained from one of his police contacts back in England.

Pointing a finger at the schematic diagram of the compound he elaborated on his plan. Alan was to infiltrate the compound, under cover of darkness, and then reconnoiter each of the buildings inside. His primary goal was to identify Commandant Joubert's laboratory but if he was unsuccessful the first time, he would return again the following night and try again. It was vital he locate the site to assess its vulnerability.

Chris had protested that he too should share in the nighttime surveillance, but to no avail. Alan had insisted that he was the expert at this kind of work and so only he would carry it out.

On the first night in Johannesburg Alan had left the hotel late at night and had spent four hours crawling around the compound. He had noted that the Institute was not heavily guarded, except for a small laboratory set back from all the other buildings. He had also noted that that a reserved parking spot, marked 'Commandant Joubert...Laboratory Director', was close to the building he had initially focused on.

Over breakfast the next day at the hotel Alan had detailed his findings to Linda and Chris and they had both agreed that the building he had pin pointed was the logical choice. On his next night's vigil, their suspicions were confirmed.

At midnight, a sleek black Mercedes had suddenly entered the compound and parked in the spot reserved for Commandant Joubert. Alan had watched the dark silhouette of a tall shadowy figure, striding from the car and entering the building he had been watching.

Alan had even heard Joubert speak to the guard inside and he had seen a light go on at the rear of the building. What luck, thought Alan, the window of Joubert's office faced directly out onto the rear of the compound. He had determined the previous night that the window was made of ordinary glass, protected by horizontal metal bars.

Taking a chance that Joubert would now be inside his office for a while, Alan had crawled on his belly to the side of the Mercedes and had examined the vehicle more closely.

He had sensed that the car sat heavily on the ground and this had made him want to examine it closer. Using a small infrared torch that he had brought with him, he had seen that the car was heavily armor plated…it also had very unusual tires. They were thick and heavy, much more than he would have expected, even for a steel reinforced vehicle of this type.

Before returning to his concealed position in the long grass, Alan had taken a small silver object from his pocket and had fixed it to the underneath of the car. The little device had made only a slight sound as its magnetic case had clamped on to the steel chassis. Just as Alan had completed this task, he had had heard a noise at the laboratory door and fearing that Joubert was about to appear; he had scrabbled back to his hide.

It had not been Joubert, instead the light of a match at the door had revealed the face of a guard who had just stepped outside for a smoke. As Alan had watched the big hulking man puffing his cigarette in the darkness, he had heard another noise…from a different direction.

Shrinking back still flatter against the ground Alan had spotted a young black girl, about twelve years old, who was picking her way quietly past the building. The guard had seen her too and he had extinguished his cigarette and shrunk back into his secret hiding place.

What happened next had made Alan's blood boil. As the young girl had passed innocently by the spot where the guard had been hiding, he had suddenly leapt out, grabbed her by the hair and dragged her behind the bushes. The terrified girl had whimpered in vain as the powerful brute had put his hand over her mouth, ripped of her clothes and forced her to the ground.

Alan had found it almost impossible to restrain himself from rushing over to prevent the girl from being ravaged. The little girl was obviously being violently raped and the sound of the blows and her cries of anguish left no doubts regarding the brutality of her attacker. Alan had laid there feeling helpless.

Joubert was still inside the building and any attempt at rescuing the poor girl would have blown his cover and jeopardized the mission.

As he lay there, forced to listen to the attack, Alan had bitten his lip until it had bled…And had sworn to deal with the cowardly beast at the very next opportunity.

Once the guard had satisfied his lust, he had left the still weeping girl in the bushes and gone back inside, Alan had been about to crawl over to help her when Joubert had suddenly appeared at the door. By the time Joubert had walked back to his car and driven out the gates of the Institute another few precious minutes had passed and before Alan was able to reach the bushes where the poor girl had been ravished…she had gone. All that remained was some blood stained clothing, the sight of which reaffirmed Alan's resolve to dispose of the inhuman beast that had committed this heinous act.

Alan had not mentioned the incident to Chris and Linda when he had got back to the hotel. He had only related his findings regarding the precise location of Joubert's nest. They could not understand why Alan had not been more excited, especially when he had described how easy it was going to be to get his explosive charge into the laboratory.

As Alan had gone on to outline the rest of his plan they had mistaken his sullen manner for fatigue, which was understandable after his two long nights of surveillance. Alan was not a man to stay down for long though and he had brightened considerably when he outlined his plan of action for the following day.

'Chris and I will penetrate the outer perimeter of the compound tomorrow morning', he said resolutely.

Linda looked frightened as she voiced her own fears.

'Isn't that too dangerous during daytime, Uncle?'

'No Linda, the place is poorly guarded and we can get to Joubert's laboratory from the rear of the compound. We need to go during normal working hours…so that you can get his home address from the receptionist.'

'What do I do while all this is going on,' said Chris.

'You Chris will keep watch on the sidelines and warn us of any trouble…as I am placing my explosives and as Linda is probing for the information that we need'

'The whole operation must not take more than 30 minutes,' Alan went on sternly.

'But why must Linda be the one at risk entering the main building?', questioned Chris.

'Because that will arouse less suspicion…Linda has a South African accent…you my friend Chris. Do not.'

164

Chris and Linda could see that Alan had made up his mind and there was no room for further discussion.

'Now let's get some sleep, we have big day ahead of us', Alan concluded. The tone of his voice made it clear that the planning stage was over…and the action stage was now about to begin.

Chapter 27:

Retribution

The following day Alan, Chris and Linda had an early breakfast at the Balalaika hotel before heading out to the Institute of Medical Research in Reitfontein. It was the first time Chris and Linda had seen it but it looked just like any other research facility back home.

Driving to the rear of the compound Alan and Chris got out of the car and Linda moved into the drivers seat so that she could drive the car to the main gate. Alan had fabricated a plastic identification badge for Linda, one that falsely declared her to be 'Dr. Antoinette Smuts, Virology Department, S.A.I.M.R. A crest of the South African Institute of Medical Research, which Alan had forged at the bottom of the badge, was an added bonus, and would pass even the closest scrutiny.

The ruse worked, for after clipping the badge to her jacket, Linda passed through the main gate without hassle, the security guard even saluted as she drove by. Parking her car in the main lot at the front of the building, Linda strolled through the glass doors and followed her nose to the doctor's lounge. She had reviewed the task that Alan had set her...now she followed her own instincts.

The doctor's lounge was empty but several discarded white coats were hung on coat hooks and two of them had stethoscopes dangling openly from their pockets. How fortunate Linda thought, as she donned one of the jackets and hung the stethoscope around her neck. The universally accepted badge of office that all doctors carried would add credibility to her deception.

Transferring the identification badge from her street clothes to the white coat she was now wearing, she sauntered out of the empty lounge and headed for the central reception area. The receptionist at the desk had no reason to doubt the authenticity of the doctor in front of her, or to question the reason for her inquiry.

'I'm looking for Dr. Joubert, do you know if he is in?'

'No doctor, the Commandant is not at work today, I believe he is at home taking some vacation.'

'I have an urgent case to discuss with him, do you have his number…or his address', Linda added innocently.

'No, I'm sorry doctor, he has an unlisted number but I know he lives in Bryanston…On Phillipa Avenue I believe'. The receptionist had heard through her internal grapevine about the wild parties that Commandant Joubert often held at his home, she also resented the fact that her advanced and plump figure had precluded her from ever getting an invitation. Linda smiled as she thanked the receptionist for her help and walked back to the doctor's lounge where she returned the borrowed coat and turned to leave.

Suddenly she remembered the identification tag and she quickly removed it from the discarded white coat and put it back on to her own jacket. Her clever ploy had worked perfectly but she must not mess it all up now by leaving any incriminating evidence behind.

Linda returned to her car and as she drove back out through the main gates, she was grinning, she had successfully accomplished the dangerous assignment she had been given…now it was up to Alan and Chris.

Meanwhile Alan and Chris had already breached the back fence and were now crouched in the long grass that surrounded Commandant Joubert's laboratory. Leaving Chris behind, Alan edged his way to the outer window of Joubert's office and cut a small hole in the glass between the vertical bars. The glasscutter he had brought with him was new and it made no sound as it circumscribed the glass before Alan extracted the perfect circle and placed it quietly on the ground.

Just then Alan heard a cough from inside the building, as the guard inside cleared his throat and he instinctively knew it was the guard that had been on duty the night before. Alan smiled as he made a slight modification to his plan. Purposely and with great deliberation he took a second silver tube from his pocket and wired it to the bomb that he had prepared.

His earlier calculations had already confirmed that the explosive device he had made, was powerful enough to destroy Joubert's laboratory and all its contents…Now he was adding just a little extra. The brutal violation of the young girl last night would not go unpunished. With no concern for the occupant inside, Alan set the timer and dropped the deadly package quietly through the window.

Alan knew that he did not have long to get clear so he briskly moved back to where Chris was hiding and motioned that he should follow him back over the fence.

Chris did not have to ask Alan how it had gone, the look on his face said it all, and as they skirted the fence, they were relieved to see Linda in the car waiting for them. As they all drove off together Alan asked his niece how she had fared and she beamed as she told him how she had accomplished her task.

'Very resourceful of you Linda, I knew I could count on you to get his location, now let's be on our way.'

As Alan slipped into the driver's seat, Linda slid over to the other side and took out a map of Johannesburg.

'Our destination is Bryanston in the northern suburbs, I'll give you more specific directions as we get closer to our target,' said Linda, warming to the adventure.

They were only halfway to their destination when a powerful explosion rocked the compound of the South African Institute of Medical Research. The blast was so powerful that even at this distance they all heard it…

The fireball that followed the explosion also blew the guard inside through the roof and his burned and broken body was blasted twenty feet into the air, before it landed in a heap of mangled flesh and shattered bones.

'My God, that was loud,' cried Linda.

'Yes I must apologies to my suppliers back home, I seem to have used a little more explosive than I really needed,' Alan replied, with a twinkle in his eye.

The trio had heard the explosion but Commandant Joubert had not, so he was still unaware of the death and destruction back at his laboratory…He was also unaware that three determined pursuers were fast closing in on him.

As Linda consulted her map, she called out directions as the car screeched around the winding roads and avenues of Bryanston. When they finally reached Phillipa Avenue, Alan slowed the car down to a crawl.

As they scanned each driveway carefully and searched for Joubert's distinctive limousine, they had almost reached the bottom of

Phillipa Avenue when a black Mercedes burst from a driveway and almost forced them off the road as it went by.

'It's Joubert,', its him yelled Linda excitedly 'Turn the car around now she screamed and get after him'.

Chapter 28:

The Hawk Escapes

Ten minutes earlier Commandant Joubert had been standing by his pool contemplating his next, and probably last dip in the pool.

He had already packed his briefcase and locked all its valuable contents in the boot of his car, but although he was eager to get started on his trip, he wanted to have one final swim in the cool waters of his sparkling pool. As he reached into the cabana for a towel the little red light on the wall went out...Right before his eyes.

His mind contorted as he tried to puzzle out why. The red light had glowed constantly for the past two weeks, ever since the handyman had messed with its contents and started the death and destruction that had followed. Joubert had not bothered to cancel out the warning light, even on his last visit to the laboratory; there had seemed no point. The constant red glow also served as a reminder to him of the coming Armageddon.

Joubert re-examined the circuitry in his mind...the light went on when someone opened the refrigerator...It only went off when Joubert reset the secret switch...It had stayed on because he had not reset it...So why had it gone off now?

Just then two distinct solutions dawned on him. Either some unauthorized person had discovered his secret micro switch...Or all the emergency power circuits at his laboratory had failed.

Fearing for the safety of his precious genetic aberrations Joubert raced inside to the phone and dialed the main switchboard at the Institute. The voice at the other end of the line was hysterical.

'Oh Commandant, there's been a terrible explosion...your laboratory was destroyed and one of your men was killed.' Joubert knew immediately what had happened.

Despite his orders, his pursuers had somehow slipped into the country and had destroyed the triggers of death that he was about to unleash on Africa...but not quite. How ironic Joubert thought, they had done to him what he had done to them but he still had a few surprises left up his sleeve.

Dressing as he ran, Joubert raced to the kitchen freezer and grabbed the six precious samples that he had cunningly stored there earlier. Then he bolted out to his car and stomped so hard on the accelerator that he skidded on the driveway, before bursting out the gate.

In his rush to leave he almost collided with a car that was coming toward him from the opposite direction. Speeding off to his new destination, he had no time to see who the occupants of the Peugeot were…Three people, who were now watching him, disappear in a cloud of dust.

Once down the avenue and out onto the open motorway, Joubert floored the accelerator and the powerful limousine surged forward like a bullet out of a gun. His disdain for the speed limit was only surpassed by his determination to reach his destination.

His focus on the road ahead also made him unaware of the white Peugeot that was now trying desperately to keep up with the speeding Mercedes ahead of it.

'Faster, Faster, he's getting away,' screeched Linda.

'I'm going as fast as I can but don't worry Linda, just get that black gizmo from the glove box and turn it on'

Linda snapped open the glove box and pulled out a small black oblong box. As she pressed a button on its base, a lid popped open to reveal a miniature screen, which immediately lit up as a dancing green dot at the top of the viewer, announced its presence by making a loud pinging noise. Linda almost dropped the device in surprise and looked at Alan for an explanation.

'Don't worry Linda, it wont bite you, its a direction finder, the green dot will tell us where Joubert is.'

'How the heck does it do that?' asked Chris.

'Because I placed a very tiny transmitter under the chassis of Joubert's car last night, this detector will now tell us where he is…Or at least I think it will'.

'What do you mean you think it will, you don't sound very confident', demanded Chris brashly.

'Well it only has a range of five miles, but it should at least tell us if he turns off the road up ahead and then hides waiting for us, somewhere in the bush.

'Linda consulted her map and then exclaimed loudly. 'I don't think he will Uncle Alan, the only place up ahead is Soweto, I doubt that he will turn in there'.

As if to refute Linda's speculations the Mercedes up ahead appeared to slow down as it drew closer to the sprawling black township. Inside the Mercedes, Joubert made a snap decision. Now was the time to start his black plague...And this was just the right place to trigger it.

Pressing a switch, he activated the electric windows and as the glass slid smoothly and silently open, the smell of wood smoke and rotting garbage came rushing into the car. Joubert wrinkled his nose in disgust as he reached into his shirt pocket and pulled out one of the tubes of blood. Smiling he visualized the satanic virus swirling around in their cellular nutrient as he flung the tube through the car window.

Joubert did not hear the crackle of breaking glass as the airborne tube struck a concrete wall at the side of the road but he knew it would not be long before the ubiquitous mosquitoes found the blood feast that he had sent them.

He was still driving alongside the sprawling township, despite his great speed and realizing just how large it was, he decided to give it yet another dose.

It was rumored that Soweto had a population of over two million people, all of them black...Certainly worth another one of his precious tubes, he now decided.

He would still have four more tubes left...more than enough for any contingency that might lie up ahead. As he carefully extracted the second tube of death from his pocket he suddenly spotted the Peugeot that was now growing ever larger in his rear view mirror.

Joubert never forgot a face...or a car, and he immediately recognized it as the one that he had narrowly missed as he had raced from his driveway.

Certain that he was now being followed, Joubert closed the windows and flicked on the turbocharger of his powerful car and the Peugeot was left far behind. Joubert had been interrupted at his work and he snarled as he replaced the tube in his pocket...Then he leered again...He would just have to give the township its second dose...On his way back.

A mile back his pursuers were fuelled by speculation. 'I'm sure that when he slowed down just now, he threw something out of the window,' insisted Alan.

'My God, I hope not…that means he may still have more of his hellish creation with him…and we didn't destroy it all in his laboratory', shrieked Linda.

'Now lets not panic', said Alan, trying to pacify them.

To take her mind off her worst fears Linda again consulted the map on her knee and tried to figure out where Joubert could be headed She prayed to herself that Joubert was not still trying to trigger his deadly plague and that he was merely trying to escape.

'Where are we headed,' queried Alan, noting that Linda was now studying the map much more intently.

'According to the map there's nothing up ahead of any significance…Not until we reach the border…The border that separates South Africa from Zimbabwe', Alan frowned at her reply and Linda saw it immediately.

'What's wrong Alan, what have you just thought of'.

'That's the trouble Linda, I'm afraid that I didn't think, not of everything anyway', said Alan blushing.

He went on to explain that he had not thought to get them a visa for entry into Zimbabwe, as he had not thought they would be going there. Now he went on to point out that by the time they lied, or bribed a way past the border guards, Joubert would be long gone.

Rather than risk paying an unauthorized visit to South Africa's Northern neighbor, Alan fell silent as he hatched another plan, then he brought the car to a sudden halt.

'Why are we stopping, we can't give up now,'

'I'm not giving up, I'm going to stay here and lie in wait until that madman comes back down this road'.

'But how do you know that he will come back this way. Or just how long that will be', pleaded Linda.

'My gut feel tells me that he will come back this way Linda and I will wait for him for as long as it takes'.

Linda again consulted her map.

Well the odds are pretty good Uncle Alan, this is the only main road that connects South Africa to Zimbabwe, all the other side roads are just dirt tracks'.

As they all piled out of the car, Linda could not help noticing the stark beauty of their new surroundings.

Although the veldt around them was brown and uncultivated the road they stood on overlooked a deep ravine. Beyond the narrow grass verge on the right hand side of the road, the rough craggy rocks descended almost vertically to a clear crystal stream that twinkled far below them. At the other side of the road the grass verge was wider and the long grass there was interspersed with thorn bushes and wild roses.

Linda and Chris went and peered over the cliff, at the silver river below, but Alan called out a stern warning.

'Hey, don't you kids go too near the edge, it's a long way down to the bottom and I don't fancy climbing down to recover your bodies'.Heeding his warning, they both stepped back from the edge and stood watching as Alan crossed to the other side of the road and started tromping through the long grass...obviously looking for the perfect spot for an ambush.

Returning to the car, Alan retrieved the direction finder from the front seat and then unlocked the boot and got out his gun. His next instructions were clear and concise. Linda and Chris were to hide in the grass, a way back from the parked car, and Alan would conceal himself in the long grass on the other side of the road.

He explained that this was an old terrorist trick, their parked car would be a momentary distraction to an oncoming driver, which would give Alan enough time to rise from the grass and get off a couple of shots.

'You will have to be very lucky to even get off two shots if Joubert is travelling fast', cautioned Chris.

'I'm hoping that one shot with this special weapon will do the trick Chris...It may be the only chance I get'

'What will we do then,' said Linda, wondering if she would still be able to perform her duties as a doctor if the patient that ended up lying in the road was Joubert.

'We will worry about that when the time comes', replied Alan, immediately sensing her inner conflict.

Alan knew that if the weapon in his hands met up to his expectations, and he felt confident that it would, they would have absolutely nothing to worry about.

'Alright you two, now lets all be ready to intercept this crazy Commandant when he returns,' concluded Alan, as he walked back across the road and took up his new position in the long grass.

Chapter 29:

The Zimbabwe Connection

As Alan, Chris and Linda were settling down for their long vigil, Joubert was burning up the miles as he headed further North toward the border. He had come to terms with the fact that he was being followed but the fact that his three pursuers had entered the country under the noses of his agents, rankled him. He would deal with his ineffectual subordinates later.

Joubert still did not know how they had found his home so quickly, but that was of no consequence now, he did not plan to return there anyway. The only items he needed were all with him in the car, either tucked snugly in his shirt pocket, or safely locked in the boot of the car. Now the only other items he needed to start his red plague were at his destination ahead.

As he reached the South African border post Joubert waved his hand at the guard who came out to salute him…and drove straight through.

The white guard knew Joubert well; he had only once made the mistake of stopping him and he did not want to experience the wrath of his superior officer again. At the next border post Joubert did not even slow down. The black guard at the Zimbabwe border crossing had seen this car and its driver many times before and he had always been paid well to let him through without hindrance.

As he waved to Joubert, his face cracked into a toothless grin as the dust swirled around his ragged old army coat. He was not paid much to watch this quiet outpost and he knew better than to delay the man whose bribes helped feed his ten hungry children.

Once across the Zimbabwe border Joubert continued to ignore the speed limit except this time there was none posted. Black workmen under the supervision of white Rhodesian foremen had built the road system here long ago, so it had been well constructed. The ongoing maintenance and the upkeep of the road signs…that was another matter. Since Zimbabwe had gained its independence and been governed by Mugabe and his tribal henchmen, there had been much deterioration and decay.

As Joubert reached the outskirts of Bulawayo he had to swerve wildly, to avoid an abandoned truck that stood rusting in the middle of the road, but he only slowed slightly as he drove into town. He picked up speed again as he drove through the wide boulevards of the downtown area, remnants of days gone by, when wagons drawn by teams of oxen had thundered through. Finally he came to a stop outside a whitewashed building at the northern edge of town.

It was the only concrete building in this area, the rest of its neighbors was ramshackle buildings, with rusty corrugated tin roofs atop their dilapidated stonewalls. A sign outside proclaimed proudly that this was the 'Zimbabwe Institute of Malaria Research'.

Unlocking the boot, he took out his precious briefcase and walked around to the side entrance, where he passed the only other car in the car park...a dusty jeep. As he turned the corner of the building he heard a windsock flapping in the breeze and saw the little plane parked on the grassy airstrip nearby. The Cessna plane stood out proudly from its humble surrounding and its silver crop spraying equipment and wing tip fuel tanks were new and shiny.

As Joubert reached the side door, the sound of his boots on the concrete walkway, alerted the solitary occupant inside, and he came out to meet Joubert.

Major Pieterse was a stout man of good Afrikaner stock and he was one of the few men that Commandant Joubert trusted absolutely. He had fought alongside Joubert in both Angola and Mozambique and he had always served the Commandant with unswerving loyalty and obedience. Unlike Joubert, the major was not a scientist but he had been an able and compliant student over the past few months.

Pieterse had absorbed all the technical information that Joubert had taught him...But only the information that Joubert had wanted him to know. More importantly Pieterse had always done as he was told and never questioned the reasoning behind the bizarre requests. Although the major held out his hand in greeting, he still stood rigidly to attention.

'Ja, it is very good to see you Commandant, I heard about the explosion on the radio and I was expecting you...was everything in your laboratory destroyed?'

'Nothing of any consequence Major Pieterse,'

Joubert still addressed the officer formally, even though he had known the man for many years. Joubert had never shown any loyalty or friendship to any of his subordinates, he preferred to keep his underlings at a distance…To him they were all expendable.

'Has the time come, Sir., Pieterse asked.

'Yes Major, the time has come for Operation Red Plague to begin at last'.

Briefcase in hand, Joubert fell in step behind the major and then followed him into the building. Once inside Pieterse stood off to one side as Joubert gazed round the room with a smile of satisfaction on his face. The room was an exact replica of Joubert's laboratory in Johannesburg, even down to the equipment on the bench and the glass case of mosquitoes in one corner. Joubert put his briefcase on the table and took out a map.

Pieterse could see that it was a continental map of Africa that identified each of the independent countries in bright colors, the major cities were also highlighted.

'You will study the targets on the map Major, while I compare the viability of your virus with the samples that I have brought with me. Pieterse puffed up with pride at the reference to 'his virus', he had only grown them as instructed but he held off on any self congratulations until he knew that he had met all the Commandant's demanding requirements.

Joubert crossed over to the refrigerator, took out four samples from the rows and rows of crimson tubes stored inside and placed them in a stainless steel rack. Returning to a bench he placed them in front of the gene machine and then took four of the five tubes from his shirt pocket and also added these to the metal rack.

For the next fifteen minutes Joubert carried out his analysis while Pieterse studied the map and calculated the distance between each city that Joubert had targeted. Kinshasa…Lagos…Bemako…he would refuel at Dakar on the West coast before flying on to his next set of targets…Kampala…Mogadishu…Nairobi…Harare.

Pieterse paused now, Harare was his last target and none of the major cities in South Africa were targeted. This made sense to him, having Harare as the last target gave him time to fly back to Bulawayo and destroy both the laboratory and the plane. Then he would drive his jeep South…to safety.

South Africa had also not been targeted, that was his secure refuge from the coming storm, and the place where his Commandant would ensure he was safe...Or so he thought. By the time Major Pieterse had completed his calculations, his Commandant had also finished his genetic evaluation.

'Congratulations Pieterse, the virus mutations you have cultured are duplicates of my control material...and you have propagated enough to meet our needs. 'Pieterse did not always understand all the terminology that his boss used, but he knew that his commander was pleased as Joubert was smiling now as he reached into his briefcase and took one of the two remaining items from his bag.

'You have done well I can now give you a reward.'

The commandant took the small vial of pale yellow serum that he had taken from his briefcase and passed it gingerly to the major, as if it contained something of great value. To Pieterse it did, and as he took it from Joubert he very carefully slipped it into his pocket.

It was the vaccine that his Commandant had promised him...The vaccine that was going to protect him from the devastation that Joubert was about to unleash...A vaccine that would make his body resistant to the plague they had concocted...Pieterse had no way of knowing otherwise.

Joubert almost felt a twinge of guilt as he saw how trustingly his gullible assistant had just taken the bait. The yellow tube did not contain vaccine...There was no vaccine on earth that could possibly protect the man...any man, from Joubert's unique and deadly hybrid.

Piling deception upon deception the Commandant issued his next and final order.

'Do not inject yourself with the vaccine for 48 hours, the serum was freshly made this morning and will only be fully effective in two days...At the completion of your successful mission...Joubert added craftily.

Although they had discussed it before, Joubert wanted to be sure that his co-conspirator followed this last command to the letter...Joubert wanted no loose ends. The serum he had given Pieterse contained no vaccine, just a squirt of plasma...And enough insulin to kill a horse.

Returning to his car Joubert hung around long enough to watch Pieterse load all the virus into the crop spraying reservoirs on the plane, and until the little Cessna bumped off down the grassy runway. Pieterse waved to him from the cockpit, as he banked the plane and headed for his first target...Kinshasa. Joubert silently mouthed the words 'good luck and safe journey', but even he had to grit his teeth as these last selfish words flowed from his mouth.

Before driving away he reached into his briefcase and took out the one item remaining, an airline ticket, which he slipped into his inside pocket, alongside his wallet. The fat wallet held enough money for the many months ahead...the ticket was Joubert's only way of securing his own immunity.

As Joubert drove away he felt no remorse for the fate of his obedient and trusting servant. In 48 hours he would have served his purpose...and he would be of no further use to him.

Pieterse would at least die quickly and painlessly and for the ruthless Joubert that was as good as it got.

Chapter 30:

Return from Bulawayo

The Zimbabwe laboratory and his doomed accomplice were soon left long behind him as Joubert raced his car South to the border, on his way back to Johannesburg. There was no traffic on the road and the scenery was dull so Joubert let his mind wander as his car gobbled up the miles to its final destination…Jan Smuts airport.

Pieterse would be over his first target…Kinshasa, he would be flying low, ready to drop his deadly cargo over the first of his densely populated targets. The glass tubes that he scattered would shatter on impact with the ground, but their microscopic contents would not be damaged…they were indestructible.

There were many strains of virus in the world and most of them had survived for thousands of years, some of them even predated mankind.

The ones being dispersed now were designer strains, made by Joubert to his exact satanic specifications. Each one was genetically engineered to be highly infectious, each would remain viable for many days. The little glass tubes that parachuted them to the ground would shatter on impact but the millions of viral agents inside would soldier on regardless of any man made obstacles. They would stay viable inside the remnants of their bloody soup until the squadrons of mosquitoes came zooming down to helicopter them on to other unsuspecting hosts. These hosts, many of them human, would then infect other mosquitoes before their hosts died. Then the ever-growing pool of infected mosquitoes would carry Joubert's deadly blood plague to all four corners of Africa.

Thoughts of the coming red plague made Joubert tap his inner pocket for reassurance. The airline ticket was safely inside his jacket pocket and it guaranteed him first class transportation from Johannesburg to Australia. Joubert had avoided telling Major Pieterse that it had been impossible to formulate a protective vaccine, even a brilliant virologist like Joubert had been unable to do it. Nor could any one else, not before the disease had run its course that was the whole point of his plan. The only hope was escape, to the distant

181

safety of the world's most southerly continent and by the time the plague struck, Joubert would be sipping a cold beer in Australia.

In two or three years, when the plague had burned itself out, by killing all its hosts, Joubert would return to South Africa. Then he would take his rightful place as an elite member of the new Republic of South Africa. As Joubert drove his reasoning became even more outlandish.

If all went according to his plan, once the plague had died out, new migrants would pour into Africa. White settlers from Israel, America, and Europe, to repopulate and rejuvenate his decimated country...Then Joubert would return to the predominantly white...United States of Africa.

Joubert's self-aggrandizing thoughts were interrupted as cool scientific logic now cut in and he did the calculations in his mind, the ones he had done so many times before. South Africa had a total population of 50 million...45 million black...5 million white. His red plague would have a mortality rate of 90 % and so for every 40 black people that died, only 4 white people would perish.

These were acceptable odds to Joubert if it led to the resurrection of white domination in his country. His calculations were as flawed as his twisted paranoia. These thoughts continued to bounce around Joubert's head as his car drew ever closer to the three people that were patiently lying in wait for him. As the Mercedes came within 5 miles of the waiting ambush, a little green blip danced on Alan Cobb's tracking screen and sang out its bleep of warning. Alan waved to his two friends across the road and motioned them to hide in the grass as he took the gun from his knee. He checked the weapon was loaded and ready for action.

Alan Cobb was about to find out if the pistol would stop the oncoming, heavily armor plated Mercedes. The weapon was a 'Belgian Five-seven N', reputed to be the most powerful handgun in the world, a weapon that could shoot a hole through a tree trunk, a brick wall, or even punch a hole through 48 layers of Kevlar armor.

The Mercedes traveled another 4 miles before Joubert saw the familiar Peugeot, parked by the grass verge, ahead of him. Joubert did not react until he was within 200 yards of the waiting vehicle, then his sense of omnipotence cut in and he pressed the accelerator. What did these poor fools think they were going to do...how could they ever stop him now?.He felt secure in the knowledge that his Mercedes was bulletproof and heavily armor plated underneath…Even a land mine would do little damage to his bombproof vehicle.

As he drew closer…Now within 50 yards…he contemplated the idea of ramming the parked Peugeot at full speed and shunting it over the steep ravine. He resisted the urge, damage to the front of his car, even minor damage, would delay his trip to the airport.

As he drew almost alongside the Peugeot a sudden movement at the right of his field of vision, caught his attention. It was the figure of Alan Cobb rising out of the long grass, with what appeared to be a small gun in his hand…and it was pointed at his car.

Even this move did not dispel Joubert's confident grin. As the armored limousine thundered on, the crack of the pistol, followed by a much louder bang, as one of his tires burst, came too late for any evasive action.

Joubert's car pirouetted on its wounded wheel and became airborne, then it cut a swathe of destruction through the tall grass as Joubert sat helpless at the steering wheel. The car continued out of control over the steep embankment. Over and over it rolled as Joubert wrestled hopelessly with the unresponsive wheel and stomped desperately on the dead pedals. Alan ran across the road and peered over the abyss.

The car continued its slide down the mountain until it landed upside down at the bottom, its three good wheels spinning noiselessly in the quiet valley below. Alan called out frantically to the others.

'Stay where you are, I'm going to climb down and survey the damage…but it's going to take me a while'

'Let me come with you,' volunteered Chris.

'No you stay here with Linda, maybe you can get help from a passing motorist…A rope would be most helpful…Or better yet some crampons, if the motorist happens to be a mountaineer'. Despite what was happening Alan Cobb had not lost his sense of humor.

Chapter 31:

The Hawk's wings are clipped

As Alan Cobb picked his way carefully down the steep hillside, Commandant Joubert regained consciousness down below, and tried to assess his injuries. His left arm was broken and his body was firmly trapped between the twisted seat and the inflated air bag that was now pressing on his chest. He was still able to move his right arm and hand and he used this hand to check his face for cuts and abrasions. There was a large gash on his forehead and he could feel the blood trickling down his face. Considering the severity of the crash, he was not in bad shape, but he had to get free from the car before someone came to get him.

The smell of gasoline hung in the air and he rebuked himself for not buying the diesel version of the Mercedes, then he congratulated himself, the well-designed car had not caught fire. These were bizarre thoughts to be having, he was obviously concussed. Moving his one good arm, he adjusted the rear view mirror and used it to scan his upper body and chest for signs of injury. Then he saw the bright red stain spreading across his shirt.

How strange thought Joubert, the size of the bloodstain in no way equated to any pain or feelings in his chest. Joubert cackled aloud as the realization suddenly hit him. It was not his blood...it was the tube of blood he had been saving for Soweto, the one he had carefully reserved for the second dosing of the teeming black township on his way back to Johannesburg.

The glass tube in his shirt pocket must have been broken by the impact of the crash, now the blood was leaking out. His mad cackling stopped almost as soon as it started. What if the broken glass had pierced the flesh beneath his shirt...The contaminated blood now mingling with his own precious fluid...was teeming with his deadly designer virus.

Despite his fear Joubert managed to cast his mind back,...Way back...To his lectures on hematology. His recalled his study notes on fluid dynamics...No it couldn't happen...The pressure of his blood flowing out of the wound would prevent the infected blood from flowing into his tissues...yes, he was almost sure.

Joubert squirmed and wriggled as he tried to determine if the flesh on his chest had been penetrated by a shard of glass but he still felt absolutely no pain. He started to relax...Then he heard the high pitched whine. A hungry mosquito had just flown in through one of the broken windows and was now searching for food. Joubert watched in horror as the obscene creature settled on the bloody mass on his shirt and probed through it. His eyes opened wider and his gaze became transfixed on the filthy insect as it tracked its tiny bloody footprints across his chest.

The mosquito was so engrossed in its feast that its snout, its rickety legs and even its wings were now smeared with the infected blood. Joubert was now so paralyzed with fear he was unable to move a muscle.

Then the obscene insect struck and he felt a sharp pinprick of pain as its thrusting proboscis pierced through his shirt and into his skin. Joubert screamed. Then he started to hallucinate, could God be sending down his hammer of revenge on the anvil of Joubert's chest?.

A rush of adrenaline forced his fear-cramped muscles back to life and he reached for the secret compartment under the ashtray. His right hand pressed the concealed switch and he watched the tray containing his pistol slide out. Reaching for the weapon he grasped his hand around it and raised the gun resolutely to his head.

Joubert had seen too many of his human guinea pigs die in agony, after he had experimentally infected them with his evil red plague, he was not going to suffer the same fate. Pressing the snub nose of the automatic to his right temple he squeezed the trigger.

Alan Cobb had only climbed half way down the mountain when he heard the shot ring out from below...it was immediately followed by a mighty explosion. As Alan looked down he saw the pillar of smoke that was rising up to meet him and tried to scramble back up the cliff. As he climbed back up again, the smoke from the explosion finally caught up with him. For a few seconds it completely engulfed him on the hillside and bathed him with two very distinctive odors.

The most predominant one he had smelled many times before in the Persian Gulf war...it was the smell of burning gasoline. The second, but much more offensive odor. he had only smelled once before, but it was one that he hated the most. A pungent, acrid smell that Alan Cobb would never forget.

It had been during the Falkland War when a troop of men in his command had been layered with napalm…It was the disgusting smell of roasting human flesh. Alan twitched his nose with distaste as he realized what had happened at the bottom of the ravine below.

For some crazy reason Joubert had fired a gun and the flash from the percussion cap, as the firing pin had pierced it, had ignited the gasoline vapors in the car. Alan continued his upward climb, he had no need to go down now, Commandant Joubert, and the deadly cargo he had been carrying, was now burnt to a crisp. Consumed in a funeral pyre that Joubert had ironically created for himself. The symbolic cremation of a warrior…no, just the death of a mad megalomaniac, thought Alan as he reached the top.

'What the hell happened Alan?', yelled Chris.

'I have no idea, but whatever it was it is all over now, so let's all jump back in the car and be on our way'.

As they drove in triumph back to the city, Alan was sprawled in the back seat, while Linda sat in the front passenger seat, next to Chris, who was now the driver. Alan cracked open the bottle of whiskey he had brought with him and after a few stiff belts he started to sing at the top of his voice. The tune sounded much like 'John Peel', but it had very different words from the famous original song…

'There were cats on the rooftop, cats on the tiles,
cats with syphilis, cats with piles,
cats with their arseholes wreathed in smiles.
As they reveled in the joys of fornication'.

'Uncle Alan…that's disgusting, you certainly didn't learn that song on the playing fields of South Africa.'

'No Linda, its an old Royal Air Force battle song, my men always sang it after they achieved a victory…'

Chapter 32:

Return to London

Back at the Balalaika hotel in Sandton, Chris and Alan went to their rooms to get the suitcases, as Linda went to the reception desk to check them all out of the hotel. Linda also used the phone in the hotel lobby to call Jan Smuts airport and reserve three seats on the next British Airways flight to London. As Chris and Alan reappeared in the lobby with their luggage, Linda called out to them.

'Come on you guys, our flight back to London departs in ninety minutes, so we had better get a move on'.

Five minutes later they had paid their hotel bill and were all in the Peugeot and on their way to the airport. It took them only about forty minutes in the light traffic to reach the outskirts of the Jan Smuts terminal. Once there, they followed the airport signs to the car rental drop off area and returned the Peugeot to Avis. Humping their luggage onto a trolley standing nearby they raced to the ticket counter to collect their tickets. Alan pulled out his wallet and took out a credit card.

'I will pay for these, it will be my governments contribution to the successful outcome of our mission, I will add it to the cost of field-testing their new gun'.

'You had better hurry, your plane leaves in less than thirty minutes and you still have to clear customs and immigration,' said the visibly concerned ticket agent.

As they pushed the luggage cart through the departures area the customs officer had no reason to delay them. All their documents seemed to be in order and he could see they were very anxious about missing their flight. He was used to travelers leaving it to the last minute. Their flight to Heathrow left precisely on time and the plane was only half full. As soon as the seat belt signs were extinguished, they all moved to the back section of the plane and found some additional empty seats. Chris purposely selected an empty row of four seats in the center, but Alan politely excused himself and discreetly chose an empty seat, two rows further back. 'I still have a very important report to write, and I am sure you two lovebirds want to be alone for a while'.

Raising all the arm rests on the seats he had chosen, Chris grinned sheepishly as he made his comment. 'Now we two can celebrate the success of our mission Linda', he said, as he reached up into the overhead storage bin, and brought down two woolen blankets. Throwing the blankets over their outstretched bodies, he whispered…'Now we can join the mile high club'.

'Whatever do you mean Christian, whatever is the mile high club,' said Linda, feigning mock innocence. Chris threw her a cheeky grin but made no reply as he pulled up the blankets until they covered them.

Alan Cobb smiled to himself as he watched the row of seats where his two friends were now stretched out. The bank of four seats was swaying together in unison. Only Alan knew what was going on, the rest of the passengers in the plane were trying to sleep as the lights dimmed and complete darkness bathed the cabin.

It had been a long time since Alan had joined the mile high club, but he smiled to himself at the thought of it.

As Chris and Linda made love together tenderly, Linda realized that this was the very first time that she had experienced her lover without his latex protection. The sensation of uninterrupted closeness made the experience all the more intimate and special. She was not at all worried. Linda loved Chris dearly and the earlier precautions she had insisted on, stemmed from her concerns regarding an unplanned pregnancy, not from her concerns as a virologist.

Now she was not concerned at all, she knew that Chris would one day marry her and at least the world was now safe…safe for the future arrival of the fruits of their union. As Linda gave herself completely to her lover she forgot about her one remaining concern.

Why had Joubert driven over the Zimbabwe border?

Chapter 33:

Back home to America

Linda was feeling somewhat sad as she prepared to say good-bye to her favorite Uncle at Heathrow airport. Chris Foster was also sad; he had enjoyed the support and the companionship of his resourceful new friend. Alan Cobb was not one for long tearful farewells so he made his excuses...'Why don't you two pick up your boarding passes for New York while I make a couple of important phone calls.'

He could sense that they were getting apprehensive about their arrival back in New York. Chris had not mentioned it, but Alan knew that he was worried about what the authorities would say when they arrived back on American soil.

Stepping over to a nearby phone kiosk, Alan made a phone call to his superior officer in London...then he made an even longer call to a friend in Washington. Alan smiled as he put down the phone, it was fortunate that he had done some bomb disposal work for his American allies and that they still remembered him. Returning to his two friends in the departure lounge Alan hugged his friends and said his final farewells.

'Just two things before you both head for your plane...first I need you to return the two British passports that my government kindly loaned you...'

'What a shame, Alan, I was hoping you had forgotten them,' said Chris, taking them both out of his pocket and handing them reluctantly to Alan Cobb.

'And what is the other thing, Uncle Alan?'

'I have just made a phone call to a drinking buddy of mine in the FBI and he has assured me that you will have no hassles on your arrival back in America.

'Whatever did you tell him?' Linda implored.

'Patience, my young friends, all will be revealed to you when my friend meets you at Kennedy airport...He will explain everything...Now go for your plane'.

Chris and Linda looked at each other quizzically but Alan held up his hand to forestall further questions, then without another word he strode briskly away.

The flight back to New York was uneventful and somewhat of an anticlimax after all of the adventures in South Africa. Chris and Linda enjoyed the hot breakfast that was served just before they landed at New York, the food was good and they felt confident that Alan Cobb had somehow placated the authorities regarding their unlawful exit from America.

As Chris polished off the remainder of his scrambled eggs he could not help voicing his remaining concern…'I'm still curious as to what happened to Devere's dead body…the one that I know I left in my refrigerator.'

Linda grinned as she cleverly mimicked her uncles parting words. 'All will be revealed at the airport, Chris…you know my Uncle has never let us down and he obviously has some very good connections with the authorities back home.'

As soon as the jumbo had touched down at Kennedy airport, and they had disembarked, Alan and Linda made their way to the customs and immigration area.

As they drew closer to the final clearance point Linda was starting to feel somewhat apprehensive. The officer inspecting their documents showed no emotion as he entered the serial numbers of their passports into his little desktop computer. The experienced and well-trained officer did not change his expression when a special message scrolled down the front of his screen…'Detain on arrival…Contact the FBI for instructions.'

The officer rose from his chair and motioned for Chris and Linda to follow him.

'Is there a problem officer', Chris asked.

'Follow me, we need to verify your documentation', the officer's face remained deadpan as he escorted them to a small room and closed the door behind him.

Chris and Linda gazed around the stark room that was sparsely furnished with a metal table and two chairs. There were no pictures on the walls and the solitary window in the room was locked and secured by heavy iron bars. As they both waited sullenly, a door opened. A middle-aged man, with short-cropped gray hair came in. He was wearing a black suit, a white shirt and a conservative striped tie…and he looked officious. However his hand was outstretched in an obvious gesture of welcome and he smiled warmly as he spoke.

'Special Agent Paul T. Whittaker…the bureau is happy to welcome you back home safely, Dr. Foster.'

Linda wrinkled her nose, the pointed reference to Chris's title and the omission of her own rankled her. It was obvious that male chauvinism was still alive and well in the FBI but she held her tongue as Chris responded.

'It's good to be back, agent Whittaker, I believe Wing Commander Cobb has explained what happened but I do have a few questions of my own.'

Agent Whittaker smiled at the way Dr. Foster had taken the offensive and had pointedly used the title of Whittaker's old drinking buddy in England. It also amused him that these errant travelers were trying to pull some rank on him, which wasn't at all necessary.

'Rather than give you a long drawn out explanation, why don't you read my report while I go get coffee.' He then passed Chris a manila folder bearing the impressive FBI seal and immediately left the room in search of their coffee. Chris opened the manila folder and with Linda now standing at his side, they both read the report together…

To Director: Federal Bureau of Investigation.
From: Agent, Paul T. Whittaker

Ref: The Surveillance of Carel Devere,
alias assistant cultural secretary, S. A. Embassy
alias computer salesman, Techtronix, San Diego
alias President, Dade County Engineering Inc., Miami.

Carel Devere was admitted to the U.S. as assistant cultural secretary at the South African embassy in Washington, DC and was placed under surveillance on the advice of Interpol. Subject was known to them as agent of B.O.S.S. (Bureau of State Security, a quasi-official South African organization, headquartered in Pretoria and engaged in terrorism and espionage, with undercover agents in the United States and the United Kingdom).

Feb.14th 2.00 p.m. Subject identified leaving his embassy in Washington by agent Wilbur Cosman, who then followed him to his destination, where he was observed placing explosives at the rear of the forensic laboratories in downtown Atlanta. Agent Cosman, in an attempt to diffuse said explosives was injured in the blast and was unable to continue surveillance (agent Cosman now recovering at St. Mary's hospital, Atlanta, from injuries sustained above). All points alert was issued to FBI agents in Atlanta with special emphasis on medical laboratories and hospitals.

(targets of subject Devere not known at this time). Feb.14th 7.55 p.m. Subject Devere spotted outside Centre for Disease Control, Atlanta by Agent Mac Cormack who continued surveillance and called for back up. Devere met with new suspect (later identified as Dr. Christian Foster of the C.D.C. in Atlanta) at 8.00 p.m. and both entered the building together at 8.05 p.m.

Foster, came out of building alone at 8.25 p.m. and left in his car. Our agent was unable to pursue him but when back up arrived (agent Wilson) both of our agents entered the building and searched the premises. Primary subject Devere found dead in Foster's laboratory, hidden in refrigerator (cause of death...Brain hemorrhage...Nasal bone driven into subjects brain...an automatic weapon also found by the body...fired once...No trace of bullet)

Body of subject Devere bagged and transported to Washington city morgue to await further instructions. Security staff at Centre for Disease Control appraised this was a special FBI investigation and were ordered not to discuss this action with anyone. Our primary objective at this time was to await claim on body, by person or persons unknown and thus identify other BOSS agents working in America, especially those that were not yet known to us. Feb 14th 10.20 p.m. suspect Foster returned to Centre for Disease Control and remained inside for ten minutes, then left the building. Foster followed by agent Mac Cormack but was unfortunately lost in traffic. Continued surveillance on all airports, rail stations and bus depots were negative and it was assumed that suspect Foster had left the area under false papers with accomplice/ accomplices unknown.

Addendum: Following call from one of our 'friendlies' in the United Kingdom, Wing Commander Alan Cobb, R.A.F.

(see attached confidential transcript) our surveillance on Dr. Christian Foster was then terminated. <u>Important Notation </u>We are completely satisfied that Dr. Foster was not involved in any undercover plot. Also his meeting with Devere was under duress and he only killed him in self-defense. Dr. Foster's subsequent actions and unauthorized flight from the area were completely justified and he is to be commended for his subsequent actions. His accomplice, Linda De Vaal is also to be congratulated.

Respectfully Yours...
Paul T. Whittaker, FBI.

****

Chris and Linda finished reading the report just as agent Whittaker came back into the room carrying their hot coffee. Linda was smiling, not with relief that her fiancée was no longer a suspect, but at the strange jargon these undercover policemen always used.

She was also amused that her part in the whole affair had somehow been given an absolutely minor role. Just as she had suspected earlier, as far as the F.B.I. were concerned, a woman's place was in the kitchen.

'So agent Whittaker, you did suspect me.', Chris said.

'Well you did act in a suspicious manner and when you disappeared without trace...we assumed that you were also part of the plot.

'Did you know what the plot was,' cut in Linda.

'No, we had no idea, not until Alan Cobb called us and gave us the full details of the plot, then we were able to piece it all together. The death of the lab worker from Detroit, plus the explosion and death of Dr. Gorrie had made no sense to us up until that point'.

'What about the arrival of Commandant Joubert in Atlanta? Why didn't you move in on him then...He was obviously bent on killing both Linda and myself.'

'Yes I'm sorry about that…Commandant Joubert was someone that we knew nothing about. He seldom traveled outside South Africa and when he did it was always as a well-respected virologist and scientist. When the visa was issued for his visit to the Centre for Disease Control we didn't know what he was up to. It was not until the explosion in your laboratory Dr. Foster and the death of your director, that we finally locked onto him. When we checked him out further we finally caught on to his connection with BOSS.'

'So why did you cover up the death of Devere, his missing body sure made me sweat.'

'We were hoping that one of his cronies from the South African embassy would come looking for him, but we never expected his boss to come all the way from Johannesburg. We always have to act cautiously in these cases Dr. Foster, Devere was acting under cover of his bogus position at the embassy and we didn't want to risk a diplomatic incident. Up until very recently we had no idea just how devastating and far reaching Joubert's plot really was.'

Linda could not resist cutting in again. 'So what exactly did my Uncle Alan tell you?'

'That must remain confidential, as must all the other details of this whole damn affair. Our department is certainly grateful for what you have done, in fact the whole country owes you both a debt of gratitude'.

As Chris and Linda blushed with embarrassment, agent Whittaker carried on speaking. 'As you can both appreciate none of this must ever be revealed, if the newspapers or the television media ever got a hold of this information it would cause widespread panic. Even though you have averted a world wide plague we cannot give either of you any official recognition…But our State Department will find some way of thanking you…Off the record of course.'

It was obvious that the discussion was over and that no further information would be forthcoming. As Agent Whittaker turned to leave, he made his last offer.

'Would you both like an escort back to Atlanta?'.

'No that will not be necessary. We have reservations on the next flight and if you will just clear us through customs and immigration, we will be on our way,' responded Chris.

Almost as suddenly as they had been whisked into this secure little area, they were ushered back out again and were again back in the immigration and customs area.

This time their papers were quickly endorsed and approved, before agent Whittaker escorted them back to the departure gate, for the last leg of their journey. Shaking both their hands he threw one final caution.

'Now that this whole dreadful thing is over I don't want you making any more trips on false passports or using airline tickets with fake identification.'

'Don't worry agent Whittaker, we don't plan on doing it again, Joubert's plague is over and the only trip we will be making will be to Hawaii on our honeymoon…As Mr. and Mrs. Chris Foster', Chris added grinning.

Agent Whittaker smiled and Linda gripped Chris's arm lovingly as the two of them left for their plane. With their intended marriage now out in the open, they both had a lot of planning to do when they got back home to Atlanta.

Chapter 34:

The Red Plague Strikes

Chris Foster and Linda De Vaal arrived back in Atlanta completely exhausted after their long journey. They felt the adventures they had just experienced were enough to last them a lifetime, but that was not to be.

The first two days back in Atlanta they did no go to work. Chris did not have a laboratory to work in and Linda was too busy making plans for their wedding to even think about her own research.

Instead they had enjoyed checking out all the local jewelers together and by the end of their shopping spree Linda was proudly wearing a quarter carat diamond ring on her engagement finger. Chris had joked that it had cost him at least six month's salary, but then he had added that she was of course worth it.

After another week had lazily rolled by they both agreed that it was now time that they both went back to work. As Chris prepared to leave for work he heard Linda throwing up in the bathroom. It was obviously 'morning sickness' but Chris made no comment as he shouted a cheery good bye as he headed out the door.

He knew that she would tell him her exciting news in her own good time, so he proceeded on out to the garage and left Linda to follow on later in her own car. Arriving at the Centre for Disease Control, on his first day back at work, Chris saw the receptionist waving to him and went over to see what she wanted.

'The new director wants to see you in his office right away Dr. Foster '.

Chris had forgotten that the untimely death of the previous director, Dr. Levine, had created a vacancy for the top position at the Centre. He surmised that while he had been away in Africa, the assistant director, Dr. Pond, had taken over the running of the Centre for Disease Control.

Chris went instinctively to Dr. Levine's old office where he was now greeted by Dr. Alan Pond.

'Glad to have you back Dr. Foster, I have some important matters to discuss with you'.

'Congratulations on your recent promotion Dr. Pond, you can certainly count on my complete co-operation'.

'That's why I wanted to see you, my elevation to director has created an opening for my old position as assistant director...and you Chris, have been highly recommended for that position'.

'Why me? There are several others here who have more seniority than I do', Chris responded modestly.

'You must have friends in very high places Chris...the recommendation for your promotion came from the Surgeon General's office, no less'.

Chris smiled to himself. The FBI agent had been true to his word. There was to be no official recognition of his actions in South Africa so they had found another more subtle way to reward him.

'Of course I totally support your new appointment. went on Dr. Pond, and your first assignment as deputy director will be to get approval for the re-building of your old laboratory'.

Chris smiled. If his new director knew anything of what had happened in South Africa, he certainly was not saying anything about it. Chris also felt sure that he would have no trouble getting the necessary funding for a new laboratory, not if his new friends in government were approving the budget. He thanked the director and hurried on his way. He was anxious to get back to his office and start work on his latest task. He also wanted to phone Linda and tell her the news of his promotion.

Chris sat himself down at his new desk, picked up the phone and dialed the number of Linda's office. As she came on the line he knew right away that something was wrong. Before Chris could tell her his good news, she cried out.

'Oh Chris, the shit has really hit the fan, please get over here right away'.

As the phone went dead, Chris was out of his office like a shot and was racing across the campus. It was not like Linda to use foul language and he knew that she now needed him by her side.

Arriving at Linda's office he found her sitting on the floor surrounded by rolls and rolls of paper that were spilling out from her still clattering Fax machine.

'Whatever is going on Linda? ', Chris said gently.

'Chris, the world has gone mad. Messages have been pouring in from all parts of Africa, since this early morning…all reporting new outbreaks of A.I.D.'s '.

'Calm down Linda, let's work through this together'.

Chris joined her on the floor and, after putting his arms around her, began to read the messages that had obviously caused her so much distress. The first report was from South Africa, where a massive outbreak of A.I.D.'s had been identified in Soweto. The death toll there had already gone over fifty thousand.

The second report was from Nairobi. Again the death toll was in the thousands and the medical officer there reported that it was a new strain of H.I.V. One with a very high mortality rate and a level of infectivity that he had never ever seen before.

Malawi…Uganda…Mozambique, all painted a similar gruesome picture. Thousands were dead or dying of a new strain of H.I.V. Chris looked grim as he stood up and went immediately to the telephone.

Placing a call to his new director, Chris outlined the details of the crisis, and as he put down the phone, his face was ashen as he turned and came back to Linda.

'We could not have destroyed all of Joubert's virus…and we know why he drove over the Zimbabwe border.

'Chris, this is terrible, whatever can we do?'.

'All that can be done for now, has been done Linda. The director will inform the department of health and all the other relevant agencies ', Chris replied meekly.

For the next two days Chris and Linda worked night and day, trying to keep track of all the many outbreaks, but the flood of reports continued to come piling in.

Already the U.S. and European newspapers had picked up on the plague and the tabloids were having a feeding frenzy. There had been widespread panic in Africa and all the airports and rail stations had been closed in an attempt to contain the massive epidemic, but it had only made matters worse.

Then by order of the United States government, and with the complete support of the World Health Organization, all travel to and from Africa was prohibited. Only mercy flights that were able to fly over the infected areas, to drop essential medical supplies by parachute, were permitted.

As the plague continued to spread, the U.S. called an emergency meeting of the United Nations and demanded that the borders of Canada and Mexico be closed. The prime ministers from Ottawa and Mexico had agreed, and even though other countries howled in outrage, 'Fortress North America' was now a reality.

Chris and Linda continued to monitor all the outbreaks, hoping and praying that the plague could somehow be contained in the now isolated continent of Africa. However as experienced virologists they knew better, there were now over 5 million deaths in Africa and they both knew that this huge viral pool could not possibly be held back for very long.

Sooner or later the ubiquitous mosquito, or one of their infected human hosts, would find a way to break out from the quarantined continent.

The news media continued to have a field day, fomenting panic as well as reporting the latest news. When the death toll reached 10 million there were riots in many major cities around the world. Frightened citizens demanded action...something... anything...but no one knew exactly what.

The Center for Disease Control had classified the new virus M.I.V., Mosquito Immunodeficiency Virus, but the news media had given it a much more appropriate name...M.A.D...Mosquito Autoimmune Disease.

Almost a week went by before the first reports started coming in from areas outside of Africa...Barcelona...Rome...London. The plague had passed the line that the world had drawn in the sand.

Commandant Joubert's airborne troops were carrying out his deadly mission, even though he was no longer alive to witness the horror of his creation.

Millions were now dying in Europe, blacks and whites alike. Joubert's calculations had been as flawed as his demented plan for the white domination of Africa.

Chris and Linda still held out some hope.

Every available virologist and genetic scientist in the world was now working frantically to find a new vaccine, a new drug, a new anything that would stop the red plague from spreading.

When the first reports came in from two locations in South America, first Santiago and then Rio de Janeiro, they knew the creeping onslaught was getting closer.

Then an unexplained viral outbreak was reported in Acapulco and the US government moved quickly to close the U.S.-Mexican border. As it turned out this particular outbreak was not H.I.V. but the American people were in such a panic that they demanded protection from every disease that was out there.

'Fortress North America' was now reduced to Canada and the U.S. and all everyone could do was watch and wait.

To the north was the longest undefended border in the world and America's only hope was that the Canadian public health system, which was similar to the U.S. model, would somehow protect them.

As Chris and Linda watched and waited in Atlanta, hundreds of miles to the north a trapper emerged from his log cabin, in the wilds of Canada's Muskoka region, and went to tend to his neglected beaver traps. The trapper had not been outside his log cabin for many days because he had been sick. So sick that for the very first time in his life he had been forced to stay home in his bed.

He had never been ill before, but this time a strange fever had wracked his body after some mosquitos had bitten him. The old trapper had been bitten by all kinds of insects before, but this time it had somehow been different.

These mosquito bites had made him really sick.

Now he was feeling much better and as he tramped through the long grass, he ignored the swarms of mosquito's that rose up, and started to feed and his arms and legs. Neither he, nor they knew he was now a carrier of the mutated A.I.D.'s virus, the one that Rick Harvey had so innocently carried by car to this area. The tube of blood that the salesman had stolen from a hospital laboratory in Detroit was about to create a new devastation, this time in Canada.

When the first reports came in of a suspected outbreak of Red Plague in a region of Canada just north of Toronto, Chris waited for the blood samples to follow. They were being sent down by courier and he refused to panic until he had analyzed the samples and viewed the blood films from these latest victims on his electron microscope.

As soon as they arrived, Chris rushed them over to his laboratory and anxiously processed them. With Linda by his side he fed the first of the Canadian specimens into the electron microscope. They both shuddered as they recognized the familiar shape on the screen, the ugly rearing head of the cobra like virus. It was unmistakable…it was Joubert's deadly offspring.

Neither of them had any idea of how it found its way into Canada but that didn't matter any more. Chris hugged Linda tightly, then went to his telephone. Picking up the receiver he placed a call to the Public Health Laboratories in Toronto and sadly confirmed his findings.

During the time that the red plague had been wreaking havoc in Africa, the rest of the world had been allocating its resources to finding some kind of cure.

All the world's top scientists had been assigned to this priority project and given the task of developing a therapy that would slow down or stop the spread of this new disease. To date no effective medication had been found but a team of researchers at the University of Calgary had come up with a new diagnostic test to detect the presence of M.I.V. in blood. It was a specific and very sensitive test that would allow doctors everywhere to identify 'Mosquito Induced Virus' from the first day of infection.

It was certainly not a cure but it would allow scientists to study the progress of the disease much sooner and thus gain a better understanding of the unique red plague virus and its mode of action.

When the Canadian authorities were notified of their first confirmed outbreak of Red Plague in Canada, they decided to take a different approach. Rather than wait for the death toll to mount, they would screen everyone in the outbreak area, using their new detection test.

This turned out to be the first lucky twist of fate.

Immediately following the Muskoka outbreak a team of eminent virologists were dispatched to the area, with orders to screen all persons…man, woman and child, within a hundred mile radius of the outbreak.

Results from this test program were to be transmitted by Fax to the Public Health Laboratories in Toronto, with simultaneous copies also sent to the Centre for Disease Control in Atlanta.

This was the second lucky twist of fate.

By nature of his newly elevated position at the C.D.C. and his credibility with the U.S. government, Chris Foster was now an integral member of the team of scientists working on the 'Muskoka Project'.

This was the third and most significant twist of fate.

As Chris sat in his office, scanning the reports that had arrived that day from Canada, the sudden arrival of a courier at his office door, demanded his immediate attention.

It was unusual for couriers to be allowed inside the C.D.C. even urgent samples were left at the front reception desk, this one had to be especially important. Springing from his chair, Chris signed the delivery note for the courier and took possession of a package. It was a small brown parcel and its bright orange sticker screamed 'Blood Sample…Extremely Urgent'. Inside the package Chris found six vials of blood and a letter from the leader of the Muskoka virology team. Chris ripped open the long white envelope and read its contents with ever growing excitement.

To: Dr. C. Foster: Centre for Disease Control,
From: Dr. D. Scott: Chief Scientist: M.I.V. Project.

Dear Dr. Foster,
Following mass screening of all persons in the Muskoka region of Canada, for the presence of M.I.V. in their blood, we have to date detected 202 persons infected with this specific virus. To date 99.5 % of those persons found positive for M.I.V. have died, or are predicted to die within the next seven to ten days, based on clinical prognosis.

More importantly, one person (0.5 %) has survived to date, even though his initial exposure to M.I.V. is estimated to have occurred many weeks ago.

This subject (Male, Caucasian, 38 years old) is a fur trapper in this area and by nature of his work in the bush, has sustained many insect bites (including mosquito) over the past twenty years. This particular patient exhibits no clinical signs of illness despite the presence of significant levels of M.I.V. in his blood and serum. Dr.

Foster I am enclosing samples of his blood (whole blood taken by venous puncture), for your further investigation and comments.

I am sure the significance of this sample is obvious.

Sincerely
David D. Scott; M.D.; Ph.D.; F.R.C.P.
Professor of Immunology and Virology.
University of Toronto.

Chris put down the letter and forced a smile.

The last line of the letter was absurdly low key for such a significant finding, especially from such an eminent scientist. The statistical calculations regarding the number of deaths and the one survivor, was also typical of someone who was more used to working in research and academia.

Snapping back to attention, Chris took one of the blood samples from its protective package and raced across the room to his new electron microscope.

Putting on some protective rubber gloves he dispensed with any other extra precautions in his excitement to view the virus…in full living color.

Chris's hands trembled as the now familiar cobra like virus appeared on his screen, but his pulse and adrenalin level went wild as he continued to scan all the virus particles in the sample. Certainly the H.I.V. was identical in structure to the one that Joubert had genetically engineered, but their numbers were much lower than expected.

Much lower than I would have expected from a person infected with the disease several weeks ago, Chris silently mouthed to himself.

Sensing that something was significantly different, Chris excitedly increased the magnification to its full power and zoomed in on the cobra like apparitions. The smile on his face grew wider as he noted that the virus particles were indeed slightly, but most significantly different…different from all the previous M.I.V. particles that he had ever seen.

As Chris's smile widened to a broad grin, he sat back in his chair and relished his new findings. The M.I.V. particles in this sample were all coated with antibody. Chris forced himself to think through the rationale behind this latest discovery, and what it would mean to the world, before he ran to the phone and blurted out the news.

The trapper in Canada had obviously been bitten by so many insects over the past twenty-five years, that he had built up a massive and unique immune system. Even the deadly M.I.V. had been unable to overcome his high level of defense and instead had triggered the trapper's blood to produce enough high levels of antibody to overpower the virus.

As a result the virus particle had not only been reduced in numbers, but the ones remaining had been coated with a specific antibody, thereby neutralizing their infectivity. This brilliant example of human physiology had saved the trapper from certain death...now it could also be utilized to save millions of others.

The antibodies from this insignificant trapper in Canada could now be duplicated and then propagated in vast amounts so that it would be available for everyone, as an M.I.V. vaccine.

Not the ones that had died already, Chris reflected. This last thought triggered Chris into action and he jumped up and grabbed for the telephone. Within an hour of his first phone call, the universal medical community embarked on what was to be the largest vaccination program that the world had ever seen.

Immunology centers around the world were mobilized to produce the vaccine in vast quantities as the United Nations moved to dictate that the new life saving serum would be made available to everyone...at no charge. This was a unique situation but it was in response to a unique problem. Cost was insignificant in the race to bring back the worlds population from the brink of extinction.

In all the excitement and frenzied activity of the past few hours, Chris had forgotten to pass on the news to the most important person in his life...Linda De Vaal. Realizing this now, Chris reached for the phone, then he hesitated and smiled to himself.

This was far too important to relay over the phone, he would somehow contain himself until he got home and then tell Linda in person. Seeing her warm smile and then basking in the brightness of her moist but happily glittering eyes would compensate for the brief wait.

The brief wait to tell his beloved Linda that the world and their unborn child could now be saved.

As Chris sped happily through the afternoon traffic he reflected on a biblical quote he suddenly remembered.

…' and the meek shall inherit the earth'.

Chris shuddered visibly as he realized just how close the Red Plague had brought humankind to the very edge of destruction. The lowly virus that Joubert had manipulated had only just been stopped from bringing this ancient prophecy to pass…this time.

THE END

TONY WALTON

About the Author

Toney Walton was born in England and joined the R.A.F. at seventeen as an Armament Technician. Following five years of military service he began a new career as a Medical Technologist and worked in several U.K. hospitals, before emigrating to Canada.

There he worked for a diagnostic company, traveling extensively throughout North America and Africa. In his discussions with virologists he questioned why the mosquito did not transmit the virus…H.I.V.

In his book Red Plague Black Death, he develops this theory and takes it to a frightening conclusion.

He is now retired from his job and travels as Director of Hospital Programs and lives in Markham, Ontario with his South African wife Charlene and their family.